Ghosts in the Sky: Drones and the Changing Face of Power

TD Barnes

Published by TD Barnes, 2025.

GHOSTS IN THE SKY: DRONES AND THE CHANGING FACE OF POWER

First edition. May 6, 2025.

ISBN: 979-8231136568

Written by TD Barnes.

Table of Contents

Chapter 1: Origins of the Drone – From Balloons to Bugs

Long before the term "drone" became embedded in modern warfare, commerce, and recreation, the idea of unmanned flight was a persistent dream in military minds. The story of drones began not in the silicon heart of 21st-century innovation, but in the smoke-filled battlefields of the early 20th century, where the earliest unmanned craft carried crude reconnaissance payloads or served as kamikaze-style bombs. From these humble beginnings, drones have evolved into some of the most sophisticated tools of modern intelligence and warfare.

Long before the Wright brothers took flight at Kitty Hawk, the idea of using the sky for observation, intimidation, or attack captured the imaginations of military planners and inventors. The roots of unmanned aerial systems (UAS) stretch back to the mid-19th century, when balloons first entered the realm of warfare. Though crude and largely uncontrollable, these early aerial devices established a tactical precedent: gaining advantage from the air without risking a human pilot.

During the American Civil War (1861–1865), both Union and Confederate forces experimented with manned balloons for reconnaissance. The Union's Balloon Corps, led by Thaddeus S. C. Lowe, used tethered hydrogen balloons to direct artillery fire and observe enemy positions. Lowe and his team launched from mobile platforms like barges and trains, transmitting observations via telegraph—an early blend of aerial surveillance and real-time communications.

But it was in 1849, over the city of Venice, that history recorded the first known attempt at unmanned aerial bombing. The Austrian military, under the direction of artillery officer Franz von Uchatius, launched approximately 200 unmanned hot-air balloons fitted with timed explosive charges. Intended to float over Venetian defenses and drop bombs, the effort was plagued by unpredictable winds. Many of the balloons drifted back toward Austrian lines, some exploding harmlessly or dangerously close to their own troops.

Despite the failure, the operation marked a pivotal moment: the deliberate use of unmanned airborne platforms for offensive military action. It was a technological and conceptual bridge between traditional artillery and the future of aerial bombardment.

Even earlier than the Austrian balloon attack, visionaries speculated about mechanical flight and aerial weaponry. Renaissance polymath Leonardo da Vinci famously sketched flying machines and air-delivered munitions, including a design resembling a modern aerial bomb. While his work remained theoretical, it reflected a growing understanding of aerodynamics and a desire to dominate the battlefield from above.

In the 19th century, inventors and engineers began filing patents for various aerial devices—some powered, others gliding or balloon-based. These included schemes for parachute bombs, gliding torpedoes, and spring-propelled aircraft. Though most remained impractical, they foreshadowed the more viable unmanned systems to come.

Following the Civil War, European militaries took interest in aerial observation. During the Franco-Prussian War (1870–1871), the besieged city of Paris used balloons to carry messages and pigeons over enemy lines. These balloons, sometimes piloted, were also occasionally launched unmanned when escape was impossible. The Prussians responded by attempting to shoot them down—an early form of anti-aircraft warfare.

Simultaneously, theorists like Konstantin Tsiolkovsky in Russia and Octave Chanute in France were advancing ideas about controlled flight, propulsion, and remote control, which would become foundational to modern UAV theory. While their work focused primarily on manned flight, their studies in lift, stability, and control mechanisms would inform later UAV designs.

By the turn of the 20th century, several military advantages to unmanned flight had become clear:

1. Risk reduction – No pilot meant fewer casualties in risky reconnaissance or bombing missions.
2. Psychological warfare – Balloons and aerial devices unnerved enemies who had never faced attack from above.
3. Force multiplication – Unmanned devices could complement manned forces, extending range or enabling simultaneous attacks.

These strategic principles remained consistent across the evolution of drone technology. Though the tools would change—from silk balloons to composite stealth aircraft—the goal remained the same: extend reach, reduce risk, and gain dominance from above.

A variety of terms exist to describe aircraft that operate without an onboard human pilot.

The word "drone" has been in use since the early days of aviation, initially describing unmanned, remotely flown target aircraft used for naval gunfire training. Notable early examples include the Fairey Queen (1920s), the de Havilland Queen Bee (1930s), and later models such as the Airspeed Queen Wasp, Miles Queen Martinet, and eventually the GAF Jindivik. The term "drone" remains widely used today, especially in civilian contexts.

Modern autonomous drones utilize a suite of advanced technologies that enable independent operation without human input. These include cloud computing, computer vision, artificial intelligence (AI), machine learning, deep learning, and thermal imaging sensors. In the recreational space, a camera drone or aerial photography drone is typically equipped with first-person video (FPV) capabilities, autonomous flight modes, or both.

NATO UAS Classification

Class	Category	Normal Employment	Normal Operating Altitude	Normal Mission Radius	Primary Supported Commander	Example Platform
Class III (< 600 kg)	Strike/ Combat	Strategic/ National	Up to 65,000 ft MSL	Unlimited (BLOS)	Theatre	Reaper
	HALE	Strategic/ National	Up to 65,000 ft MSL	Unlimited (BLOS)	Theatre	Global Hawk
	MALE	Operational/ Theatre	Up to 45,000 ft MSL	Unlimited (BLOS)	JTF	Heron
Class II (150 kg–600 kg)	Tactical	Tactical Formation	Up to 18,000 ft AGL	200 km (LOS)	Division, Brigade	Watchkeeper
Class I (< 150 kg)	Small (>15 kg)	Tactical Unit	Up to 5,000 ft AGL	50 km (LOS)	Battalion, Regiment	Scan Eagle
	Mini (< 15 kg)	Tactical Sub-unit (manual or hand launch)	Up to 3,000 ft AGL	Up to 25 km (LOS)	Company, Platoon, Squad	Skylark
	Micro (<66 J)	Tactical Sub-unit (manual or hand launch)	Up to 200 ft AGL	Up to 5 km (LOS)	Platoon, Squad	Black Widow

A more technical term, Unmanned Aerial Vehicle (UAV), is often used in military and defense contexts. The UAV is defined as a "powered, aerial vehicle that does not carry a human operator, uses aerodynamic forces to provide lift, can fly autonomously or be remotely piloted, may be expendable or recoverable, and is capable of carrying either lethal or non-lethal payloads." Missiles are generally excluded from this category, though some loitering munitions — especially propeller-driven systems — are informally referred to as "kamikaze drones."

RPAS as a "set of configurable elements consisting of a remotely piloted aircraft, its control station, command and control links, and any other system elements required during flight operation."

As aerial technology has evolved at a rapid pace in the 21st century, so too has the language used to describe it. In keeping with modern standards of inclusivity and precision, the term *"uncrewed"* is increasingly preferred over *"unmanned"* when referring to aerial vehicles that operate without a human pilot onboard. This linguistic shift reflects not only cultural sensitivity but also the expanding diversity of applications and operators in this growing field.

Yet even amid this evolution in terminology, legal definitions remain uneven across the globe. One of the more persistent ambiguities lies in the distinction between hobbyist model aircraft and fully capable uncrewed aerial vehicles (UAVs). In the United States, the Federal Aviation Administration (FAA) has taken a broad stance, defining nearly any aerial platform without a human onboard—regardless of size, speed, or intended use—as a UAV. This inclusive definition encompasses everything from toy drones flown in backyards to sophisticated reconnaissance platforms deployed in military theaters.

The term *Uncrewed Aerial Vehicle* (UAV), or its alternate form *Remotely Piloted Aerial Vehicle* (RPAV), represents only one component of a broader and more integrated concept: the *Uncrewed Aircraft System* (UAS). A UAS includes the aircraft itself, as well as the suite of associated technologies necessary for its operation—ground control stations, satellite or radio communication links, data processing hubs, and mission-support infrastructure. In 2005, the U.S. Department of Defense (DoD) and the FAA formally adopted this system-based terminology in their *Unmanned Aircraft System Roadmap*, establishing a new foundation for how such technologies would be conceptualized, regulated, and deployed. Since then, the terminology has gained traction across global institutions. The International Civil Aviation Organization (ICAO), the United Kingdom's Civil Aviation Authority, and the European Union's SESAR Joint

Undertaking have all embraced the UAS model as the international regulatory standard.

Other closely related terms have emerged, often in parallel. These include *Unmanned Aerial Vehicle System* (UAVS), used in some military documentation, and *Remotely Piloted Aircraft System* (RPAS), a term favored by civil aviation authorities in Europe and the Asia-Pacific region. Canada, in a move toward regulatory clarity, adopted RPAS as its official designation on June 1, 2019, aligning its policy with ICAO language and distinguishing clearly between recreational drones and operational systems requiring oversight.

Beyond terminology, uncrewed aerial vehicles are categorized according to a range of technical and operational criteria—much like traditional manned aircraft. Classification systems differ slightly by jurisdiction but typically account for factors such as maximum takeoff weight, propulsion type, operational altitude, communication method, autonomy level, and mission scope. In the United States, the Department of Defense outlines five distinct groups of military UAVs, ranging from lightweight tactical craft such as the *RQ-11 Raven* to large, high-altitude systems like the *RQ-4 Global Hawk*. NATO and ICAO, meanwhile, provide harmonized frameworks intended to enable cross-border recognition and regulatory alignment, particularly important as uncrewed systems increasingly operate in both civilian airspace and multinational defense coalitions.

What began as a niche hobby of radio-controlled aircraft has evolved into a global enterprise, spanning the full spectrum of civilian innovation, military doctrine, and international regulation. Today, UAVs—more accurately referred to as uncrewed aerial vehicles—populate the skies in growing numbers, varying dramatically in size, function, range, and sophistication. As these systems multiply and diversify, understanding their classifications is not merely academic—it is essential to ensuring operational safety, regulatory consistency, and strategic coherence in an increasingly autonomous airspace.

One of the more practical ways to classify UAVs is by their operating altitude and corresponding range. This approach is often used by NATO, international aviation forums, and industry conferences such as the ParcAberporth Unmanned Systems event. At the lowest tier are hand-held

UAVs, which operate at altitudes up to 2,000 feet and have an effective range of around two kilometers. These compact drones are typically used for short-range reconnaissance, by infantry units or law enforcement personnel, and by private surveyors using commercial models like the DJI Mavic or FLIR Black Hornet.

Beyond this level are close-range UAVs, which extend their operational ceiling to 5,000 feet and can reach up to ten kilometers in distance. These platforms often serve in urban surveillance and infrastructure inspection roles. NATO-type UAVs, a more standardized category, typically operate at altitudes approaching 10,000 feet and can cover distances up to fifty kilometers. Such drones are often seen in battlefield support roles, offering extended reconnaissance to forward units.

As the classification scale continues upward, tactical UAVs reach altitudes of 18,000 feet with ranges around 160 kilometers. Systems such as the RQ-7 Shadow fall into this class, supporting military operations with persistent surveillance and targeting data. Further still are the medium altitude, long endurance (MALE) drones like the MQ-9 Reaper. These aircraft fly as high as 30,000 feet and have ranges that exceed 200 kilometers, allowing them to perform long-duration intelligence, surveillance, and strike missions across large swaths of territory.

The next step in operational capability belongs to high altitude, long endurance (HALE) platforms, such as the RQ-4 Global Hawk. Operating above 30,000 feet, these drones are capable of intercontinental surveillance missions lasting over 30 hours, providing commanders with near-real-time intelligence from thousands of miles away. They are among the most sophisticated uncrewed systems in existence, incorporating advanced sensor suites and satellite communication links.

Even higher up the scale are hypersonic UAVs, a cutting-edge class that includes platforms designed to fly at speeds exceeding Mach 5 and altitudes greater than 50,000 feet. Although still largely experimental, programs such as DARPA's HTV-2 and concept vehicles like the SR-72 reflect the future direction of UAV development, potentially enabling strike or surveillance missions that reach the edge of space in minutes. Some experimental vehicles are even being designed to reach Low Earth Orbit, blurring the line between drone and spacecraft. In the long term, the possibility of Earth-Moon transfer UAVs—so-called cislunar vehicles—has entered the realm of aerospace planning, especially in support of lunar infrastructure and exploration programs.

Specialized classifications also exist for UAVs launched from aircraft carriers. The U.S. Navy's MQ-25 Stingray, for example, incorporates a Computer-Assisted Carrier Guidance System (CACGS) that enables precision landings on moving vessels, marking a milestone in uncrewed naval aviation.

In parallel with altitude-based distinctions, UAVs are also classified by their power sources, which significantly influence flight duration, acoustic signature, environmental impact, and mission utility. Battery-powered electric drones are among the most common, offering quiet and efficient operation with relatively short flight times. Their low noise and zero-emission footprint make them particularly well-suited for urban use, wildlife monitoring, and intelligence gathering in sensitive environments.

Fuel-powered UAVs, on the other hand, use internal combustion engines—running on gasoline, diesel, or aviation fuel—to achieve longer flight times and greater payload capacity. These platforms are more complex, requiring maintenance and producing more noise, but they are indispensable for missions requiring extended range and endurance. Systems like the ScanEagle and Heron have demonstrated their value in military and commercial applications alike.

To bridge the limitations of each system, some UAVs now employ hybrid configurations, combining electric and fuel propulsion. This allows them to take off and land quietly under battery power, then cruise over long distances using internal combustion. Hybrid designs offer flexibility for multi-phase missions and are increasingly being tested for border patrol, disaster response, and environmental research.

Perhaps the most forward-looking category involves hydrogen fuel cell propulsion. Offering a combination of long endurance and minimal heat or noise signatures, hydrogen-powered UAVs show promise for both civilian and military use. Companies such as Doosan Mobility Innovation and H3 Dynamics are pushing this frontier, with test flights already demonstrating the viability of this technology. The high energy density and clean emissions of hydrogen fuel position it as a key player in the next generation of uncrewed systems.

As UAVs continue to grow in complexity and capability, these classification systems—based on altitude, range, and propulsion—are critical to understanding how such technologies are being deployed. From hand-launched scout drones to hypersonic surveillance platforms and hydrogen-powered sentinels, the skies are becoming a domain of extraordinary diversity. What was once a technical novelty has become a defining force in modern airspace—reshaping how we see, move, and respond to the world.

What began as a niche hobby of radio-controlled aircraft has evolved into a global enterprise, spanning the full spectrum of civilian innovation, military doctrine, and international regulation. Today, UAVs—more accurately referred to as uncrewed aerial vehicles—populate the skies in growing numbers, varying dramatically in size, function, range, and sophistication. As these systems multiply and diversify, understanding their classifications is not merely academic—it is essential to ensuring operational safety, regulatory consistency, and strategic coherence in an increasingly autonomous airspace.

One of the more practical ways to classify UAVs is by their operating altitude and corresponding range. This approach is often used by NATO, international aviation forums, and industry conferences such as the ParcAberporth Unmanned Systems event. At the lowest tier are hand-held UAVs, which operate at altitudes up to 2,000 feet and have an effective range of around two kilometers. These compact drones are typically used for short-range reconnaissance, by infantry units or law enforcement personnel, and by private surveyors using commercial models like the DJI Mavic or FLIR Black Hornet.

Beyond this level are close-range UAVs, which extend their operational ceiling to 5,000 feet and can reach up to ten kilometers in distance. These platforms often serve in urban surveillance and infrastructure inspection roles. NATO-type UAVs, a more standardized category, typically operate at altitudes approaching 10,000 feet and can cover distances up to fifty kilometers. Such drones are often seen in battlefield support roles, offering extended reconnaissance to forward units.

As the classification scale continues upward, tactical UAVs reach altitudes of 18,000 feet with ranges around 160 kilometers. Systems such as the RQ-7 Shadow fall into this class, supporting military operations with persistent surveillance and targeting data. Further still are the medium altitude, long endurance (MALE) drones like the MQ-9 Reaper. These aircraft fly as high as 30,000 feet and have ranges that exceed 200 kilometers, allowing them to perform long-duration intelligence, surveillance, and strike missions across large swaths of territory.

The next step in operational capability belongs to high altitude, long endurance (HALE) platforms, such as the RQ-4 Global Hawk. Operating above 30,000 feet, these drones are capable of intercontinental surveillance missions lasting over 30 hours, providing commanders with near-real-time intelligence from thousands of miles away. They are among the most sophisticated uncrewed systems in existence, incorporating advanced sensor suites and satellite communication links.

Even higher up the scale are hypersonic UAVs, a cutting-edge class that includes platforms designed to fly at speeds exceeding Mach 5 and altitudes greater than 50,000 feet. Although still largely experimental, programs such as DARPA's HTV-2 and concept vehicles like the SR-72 reflect the future direction of UAV development, potentially enabling strike or surveillance missions that reach the edge of space in minutes. Some experimental vehicles are even being designed to reach Low Earth Orbit, blurring the line between drone and spacecraft. In the long term, the possibility of Earth-Moon transfer UAVs—so-called cislunar vehicles—has entered the realm of aerospace planning, especially in support of lunar infrastructure and exploration programs.

Specialized classifications also exist for UAVs launched from aircraft carriers. The U.S. Navy's MQ-25 Stingray, for example, incorporates a Computer-Assisted Carrier Guidance System (CACGS) that enables precision landings on moving vessels, marking a milestone in uncrewed naval aviation.

In parallel with altitude-based distinctions, UAVs are also classified by their power sources, which significantly influence flight duration, acoustic signature, environmental impact, and mission utility. Battery-powered electric drones are among the most common, offering quiet and efficient operation with relatively short flight times. Their low noise and zero-emission footprint make them particularly well-suited for urban use, wildlife monitoring, and intelligence gathering in sensitive environments.

Fuel-powered UAVs, on the other hand, use internal combustion engines—running on gasoline, diesel, or aviation fuel—to achieve longer flight times and greater payload capacity. These platforms are more complex, requiring maintenance and producing more noise, but they are indispensable for missions requiring extended range and endurance. Systems like the ScanEagle and Heron have demonstrated their value in military and commercial applications alike.

To bridge the limitations of each system, some UAVs now employ hybrid configurations, combining electric and fuel propulsion. This allows them to take off and land quietly under battery power, then cruise over long distances using internal combustion. Hybrid designs offer flexibility for multi-phase missions and are increasingly being tested for border patrol, disaster response, and environmental research.

Perhaps the most forward-looking category involves hydrogen fuel cell propulsion. Offering a combination of long endurance and minimal heat or noise signatures, hydrogen-powered UAVs show promise for both civilian and military use. Companies such as Doosan Mobility Innovation and H3 Dynamics are pushing this frontier, with test flights already demonstrating the viability of this technology. The high energy density and clean emissions of hydrogen fuel position it as a key player in the next generation of uncrewed systems.

As UAVs continue to grow in complexity and capability, these classification systems—based on altitude, range, and propulsion—are critical to understanding how such technologies are being deployed. From hand-launched scout drones to hypersonic surveillance platforms and hydrogen-powered sentinels, the skies are becoming a domain of extraordinary diversity. What was once a technical novelty has become a defining force in modern airspace—reshaping how we see, move, and respond to the world.

The Dayton-Wright Airplane Company also created a pilotless aerial torpedo, timed to explode upon reaching a predetermined distance — a concept that foreshadowed modern loitering munitions.

By the 1930s, the idea of remote-controlled aircraft had gained practical traction:

Reginald Denny, a Hollywood actor and model aviation enthusiast, developed one of the first successful scaled remote-controlled aircraft in 1935. His designs contributed to the U.S. military's adoption of drone systems during World War II.

In the Soviet Union, engineers experimented with remotely piloting large aircraft, including modified Tupolev TB-1 bombers, in the late 1930s — an effort that demonstrated both ambition and early technical capability in uncrewed warfare.

As World War I engulfed Europe, the horrors of trench warfare and the strategic importance of air superiority drove militaries to explore every possible aerial advantage. While balloons and manned biplanes played critical roles in reconnaissance and light bombing, a new concept quietly emerged behind the scenes in American military laboratories: the pilotless airplane.

The idea of an autonomous flying bomb—capable of reaching a distant enemy target without a pilot onboard—was no longer the stuff of science fiction. It was becoming a practical engineering challenge. By the final year of the war, American industrial and scientific talent converged to produce one of the most audacious technological experiments of its time: the Kettering Bug.

In 1917, the U.S. Army Signal Corps enlisted Charles F. Kettering, head of research at General Motors and founder of Delco Electronics, to develop a pilotless aerial torpedo. Kettering, already renowned for innovations like the electric starter motor and ignition systems, brought industrial rigor and scientific discipline to the endeavor.

Working with a team that included Orville Wright as a consultant, Elmer Sperry (of gyroscope fame), and Carl Norden, who would later invent the famed Norden bombsight, Kettering's team set out to create a machine that could navigate itself to a target and deliver an explosive payload—all without human intervention after takeoff.

The result was the Kettering Bug, a spindly, wood-and-paper biplane with a wingspan of just over 15 feet. Powered by a 40-horsepower, two-cycle engine derived from a lawn mower motor, the Bug was designed to carry a 180-pound warhead over a range of approximately 75 miles.

Without radio control, the Bug relied on ingenious mechanical navigation. Its autopilot system used a gyroscopic stabilizer invented by Elmer Sperry to keep the craft level in flight. A barometric pressure sensor helped it maintain altitude.

For navigation, engineers calculated the engine's revolutions per minute and used a mechanical odometer linked to the propeller shaft. After a pre-calculated number of revolutions—corresponding to the desired flight distance—the Bug's engine was cut off, and the wings detached. The fuselage would then plunge nose-first, detonating on impact like a guided missile.

Test flights were conducted at McCook Field in Dayton, Ohio, and later at Wright Field. Some flew relatively straight and true; others veered off course or crashed on launch. Yet despite its rudimentary technology, the Bug's basic operational concept—autonomous flight, inertial navigation, and target strike—mirrored what modern cruise missiles achieve today with vastly more precision.

Approximately 45 Kettering Bugs were built by the time the Armistice was signed in November 1918. The war ended before any saw combat. The Army, shifting priorities in the wake of peace, ordered the program halted, and many of the test aircraft were scrapped on site. Though never deployed in battle, the Kettering Bug marked a radical innovation—a pilotless flying bomb with a preset course, decades ahead of its time.

Several key limitations prevented it from reaching operational maturity. Its guidance system, based primarily on gyroscopic stabilization and preset distance counters, suffered from inertial drift. Wind or slight miscalculations could easily throw the aircraft off its intended course. Launching the Bug also presented a formidable challenge. Each mission required a specially constructed rail system, along which the aircraft would accelerate atop a four-wheeled dolly before liftoff. This setup was logistically demanding and vulnerable to field conditions. Even when launched successfully, accuracy remained limited. The design called for the engine to cut off at a calculated point, sending the Bug into a glide or dive toward the target. But variations in speed, atmospheric conditions, and engine timing often caused the final descent to veer away from its intended mark.

Despite these flaws, the Kettering Bug laid the groundwork for a century of uncrewed weapon systems. Its legacy would later be echoed in cruise missile development, and its creators—including Charles Kettering, Elmer Sperry, and the Army Signal Corps—helped shape the intersection of aviation, automation, and warfare.

Nevertheless, the project proved a critical concept: that controlled, uncrewed aerial vehicles could be designed, built, and flown with combat utility in mind. The Kettering Bug was more than an experiment—it was a proof of principle that helped break conventional thinking about the future of warfare. In doing so, it also forged new collaborations between military leaders and the private industrial sector, setting a precedent for public-private partnerships that would define U.S. defense innovation for generations to come.

Though the Bug was ultimately shelved as a wartime curiosity, its legacy would echo across the 20th century. The machine introduced foundational ideas that would reappear in increasingly sophisticated forms: guided missiles, cruise missiles, reconnaissance drones, and high-altitude autonomous aircraft. It marked the beginning of a transformation in how militaries approached the concept of airpower—less as a manned platform and more as a system of capabilities extended through automation and remote control.

The program also brought together an extraordinary cast of innovators whose individual contributions would shape the trajectory of modern flight. Charles Kettering, who gave the Bug its name, continued his work as a major force in American industrial innovation, most notably with General Motors and DuPont. Elmer Sperry advanced the development of gyroscopic stabilization, a cornerstone of both naval and aviation control systems. Carl Norden, who had worked on stabilization systems for the Bug, would go on to revolutionize aerial bombing with the Norden bombsight, a key technology for U.S. aircrews in World War II. And Orville Wright, still a living symbol of powered flight, remained a guiding influence in military aviation circles long after his historic achievement at Kitty Hawk.

In an era defined by biplanes, fabric-covered wings, and cockpit duels over the Western Front, the Kettering Bug stood out as a technological outlier—a brief but brilliant glimpse into the drone revolution yet to come.

The end of World War I did not extinguish interest in pilotless flight. While the battlefield had yet to fully embrace autonomous weapons, militaries and engineers around the world remained intrigued by the promise of uncrewed systems. The interwar period, stretching from 1919 to the eve of World War II in 1939, witnessed foundational advances in radio control, gyroscopic stabilization, and aerial target applications—developments that would quietly lay the groundwork for the drone revolution of the modern era.

This era marked a shift in how the concept of a drone was viewed. No longer confined to isolated laboratory experiments or conceptual war machines, uncrewed aircraft began to emerge as methodical research projects with practical military utility. Perhaps the most important of these early roles was in aerial gunnery training. For the first time, remotely operated aircraft were regarded not only as speculative weapons of the future but as useful tools for preparing military personnel in the present.

In the United Kingdom, pioneering efforts were already underway before the war had even ended. In 1917, under a veil of secrecy, the Royal Flying Corps began developing what it called an "Aerial Target"—a rudimentary flying bomb controlled by wireless signal. The project was led by Professor Archibald Low, an innovative electrical engineer working for the Royal Aircraft Establishment. Low's team succeeded in building and flying a small biplane remotely, guided by radio waves—a remarkable achievement given the technological constraints of the time. Though the aircraft never reached combat deployment, the effort proved that flight control without a pilot in the cockpit was achievable over modest distances. Low's work would later earn him recognition as the father of radio guidance systems, a foundational figure in the field of remotely piloted aviation.

By the 1930s, the British military had shifted its focus from offensive uncrewed aircraft to training applications. The Royal Navy, seeking a safer and more realistic way to prepare its anti-aircraft gunners, commissioned a new radio-controlled aircraft: the de Havilland DH.82B Queen Bee. Based on the Tiger Moth trainer, the Queen Bee was modified to fly under radio control, allowing it to serve as a moving target for naval gunfire exercises. These drones were launched from both land-based airfields and naval vessels, flown remotely by ground or shipboard operators. Their stability, recoverability, and low cost made them ideal for repetitive training drills. Between 1935 and 1943, over 400 Queen Bees were constructed and deployed.

The Queen Bee was notable not only for its practical value but also for its linguistic legacy. Many historians believe that the aircraft's name inspired the broader term "drone" to describe uncrewed aerial vehicles. Like the male honeybee that performs a limited but essential function in the hive, the aerial drone was seen as a servant of its controllers—silent, obedient, and expendable.

Meanwhile, across the Atlantic, the United States observed British progress with growing interest. American military planners began investing in similar radio-controlled target systems, with much of the interwar effort focused on training anti-aircraft artillery crews. As the perceived threat of air attack grew in the 1930s, these pilotless aircraft became indispensable for simulating enemy bombers and testing new weapons systems. Though often overlooked in broader military histories, the interwar period was a critical bridge between early conceptual experiments and the battlefield drones that would emerge in the global conflicts to follow.

One of the most innovative minds to shape the development of uncrewed flight during this period was Reginald Denny, a British-born World War I pilot who found an unlikely second career as a Hollywood actor. After settling in Los Angeles, Denny remained deeply captivated by aviation, channeling his lifelong fascination into the world of remote-controlled model aircraft. In the early 1930s, he opened a model airplane shop in Hollywood, a venture that quickly caught the attention of the U.S. military. Denny began building increasingly sophisticated radio-controlled aircraft, adapting hobbyist designs into practical tools for gunnery training and experimental flight control.

This work led to the founding of the Radioplane Company, which would go on to produce the Radioplane OQ-2—the first mass-manufactured drone in American history. While the OQ-2 emerged on the eve of World War II, Denny's foundational contributions came during the interwar years, when he began converting model aircraft into full-scale aerial targets for the U.S. Army Air Corps. His efforts were made possible by recent technological breakthroughs in miniaturized electronics, especially vacuum tube radios and analog control systems. These innovations allowed command signals to be transmitted more reliably over short distances, making real-time flight corrections possible. Though the systems remained vulnerable to interference and signal loss, they represented a decisive step forward in the viability of remotely piloted aircraft.

Alongside Denny's advances, the 1920s and 1930s saw steady progress in the technologies that would later become essential to UAV programs. Gyroscopic stabilization, first explored by Elmer Sperry during World War I, was refined during this period as mechanical gyroscopes became smaller and more accurate. These devices provided much-needed stability, compensating for pitch, roll, and yaw disturbances in flight. Improvements in vacuum tube technology also played a pivotal role. By reducing size and increasing reliability, engineers were able to embed radio receivers in small airframes—opening the door for practical remote control at the scale of operational drones.

Electric actuators and servomechanisms began to replace earlier, less responsive mechanical systems. These electromechanical devices allowed control surfaces such as rudders and elevators to react to radio commands, granting remote pilots a new degree of control over aircraft movement across all three axes. At the same time, aerodynamic understanding benefited from a new scientific tool: the wind tunnel. For the first time, model aircraft were subjected to controlled airflow in laboratory settings, allowing engineers to study lift, drag, and overall stability in precise detail. These insights led to more refined airframe designs and eliminated much of the trial-and-error guesswork that had plagued earlier drone experiments.

Though few of these systems were fully operational by the late 1930s, the groundwork had been laid. What had once been the domain of enthusiasts and speculative military experiments was now transforming into a credible field of research and production. A new chapter was quietly unfolding—one in which drones would no longer be curiosities, but increasingly essential instruments of training, surveillance, and eventually, war.

Though these early developments did not immediately produce combat-ready drones, they were crucial in moving uncrewed aviation from theoretical exploration to practical implementation. Engineers were beginning to understand how to make an aircraft respond to remote commands, maintain stability in flight, and navigate along a predetermined path—all without the presence of a human pilot onboard. What had once seemed like science fiction was now becoming a matter of engineering challenge rather than impossibility.

As the 1930s unfolded and the threat of another global conflict loomed ever closer, military visionaries began to reassess the role of drones. What had started as training aids—tools to help prepare gunners for aerial combat—were now being viewed through a more strategic lens. The interwar years had revealed three critical truths. First, remote flight was indeed achievable, using combinations of radio communication and gyroscopic stabilization. Second, uncrewed aircraft were not only operationally feasible but economically advantageous. They could be built inexpensively, reused multiple times in controlled environments, and deployed without risking pilot lives. And third, perhaps most significantly, a new form of warfare was taking shape—one that did not necessarily require a human presence in the cockpit.

Though the world's attention remained fixed on conventional rearmament—tanks, bombers, infantry weapons—quiet revolutions were underway in backrooms and airfields from southern England to coastal California. The age of the drone was no longer speculative. It was imminent.

The outbreak of World War II marked the most transformative period in the history of unmanned aviation up to that point. The technological foundations laid in the interwar years were suddenly brought to bear on a global scale. For the first time, drones were not just conceived—they were mass-produced, deployed in operational theaters, and in some cases, turned into weapons. As nations mobilized for total war, two major branches of drone evolution emerged. One focused on target drones, used to train pilots and anti-aircraft crews in increasingly realistic conditions. The other pushed into darker territory—offensive drones built for destruction, used to deliver payloads, conduct suicide missions, or terrorize civilian populations.

The Second World War would not just change the world—it would irrevocably alter the role of machines in human conflict, accelerating the drone from a laboratory curiosity to a battlefield presence.

As the United States prepared for global conflict in the late 1930s, military leaders faced a pressing need: how to train thousands of anti-aircraft gunners quickly, safely, and economically. With enemy aircraft expected to dominate future battlefields, realistic gunnery training became an urgent priority. The solution arrived in the form of radio-controlled target drones—compact, pilotless aircraft that could simulate enemy bombers in live-fire exercises. No company would play a more pivotal role in this effort than Radioplane, the brainchild of Reginald Denny, a British-born World War I pilot and Hollywood actor turned aeronautics innovator.

Denny had spent much of the interwar period experimenting with radio-controlled model aircraft and had long believed in their potential as military training tools. With financial support and formal backing from the U.S. Army Air Forces, his Radioplane Company designed and manufactured the OQ-2, a small, twin-boom, propeller-driven drone that would become the first mass-produced uncrewed aerial vehicle in American history. Powered by a two-cylinder piston engine and controlled via radio from the ground or a trailing aircraft, the OQ-2 offered a simple but effective solution to the gunnery training challenge.

Launched from a portable catapult and recovered by parachute, the drone was both efficient and reusable. It could fly at speeds approaching 85 miles per hour and reach altitudes of up to 8,000 feet, convincingly mimicking the speed and flight profile of hostile aircraft. Constructed largely of wood and fabric, the OQ-2 was designed for ease of production and cost savings—qualities that allowed it to be built by the thousands.

By the end of the war, more than 15,000 drones in the OQ-series had been produced and deployed. Used by all branches of the U.S. military, they operated from both coastal installations and remote inland airfields, giving American gunners the chance to fire on moving aerial targets without endangering lives or piloted aircraft. The Radioplane drones helped prepare a generation of servicemen for the realities of aerial combat, contributing directly to the effectiveness of American air defenses.

Amid this surge in production and innovation, the Radioplane story took an unexpected cultural turn. In 1944, a U.S. Army photographer named David Conover was assigned to visit the Radioplane plant at Van Nuys Airport in California. Tasked with documenting the contributions of women on the home front, Conover began photographing factory workers—many of whom had taken on roles traditionally held by men.

Among them was a 19-year-old named Norma Jeane Dougherty. With her striking features and natural presence before the camera, she stood out immediately. Conover encouraged her to pursue modeling, a suggestion that launched a career which would eventually transform the young drone assembler into Marilyn Monroe, one of the most iconic figures of the twentieth century.

Though anecdotal, her story reflects a broader truth about the era: the war effort had permeated every corner of American society. Civilian industries, local businesses, and individuals from all walks of life were drawn into the machinery of war production. In Norma Jeane's case, the intersection of military technology and cultural transformation was as unexpected as it was enduring—another reminder that even the factories building pilotless aircraft were shaping history in ways far beyond the battlefield.

While the United States concentrated on defensive applications of uncrewed flight during World War II, Nazi Germany pursued a darker and more lethal course. In a bid to retaliate against the relentless Allied bombing of German cities and to demoralize the British population, the German military developed a revolutionary weapon: the Fieseler Fi 103, better known as the V-1 flying bomb.

The V-1 was the world's first operational cruise missile—a pilotless aircraft propelled by a pulsejet engine that emitted a loud, stuttering roar. This unnerving sound quickly earned it the nickname "buzz bomb." Measuring twenty-seven feet in length with a seventeen-foot wingspan, the V-1 could deliver an 1,870-pound warhead at speeds nearing 400 miles per hour and ranges up to 150 miles. Its guidance system was primitive but effective: a preset gyroscopic control directed the aircraft along a simple course, while a mechanical timer shut off the engine once the target distance had been reached. After the engine cut out, the bomb plunged silently toward the ground.

Between June 1944 and March 1945, more than 9,500 V-1s were launched at Britain, the vast majority aimed at London. The effect was both physically and psychologically devastating. Over 6,000 civilians were killed, tens of thousands injured, and large swaths of the capital were reduced to rubble. But perhaps the most harrowing consequence was the psychological toll. Residents quickly learned that the terrifying drone of the V-1's engine was not the most dangerous part of the attack—it was the silence that followed. When the buzz abruptly stopped, it meant the bomb was falling.

To combat the V-1 threat, British forces deployed a layered defense network that included anti-aircraft artillery, a curtain of barrage balloons, and an aggressive aerial interception campaign. Fighter pilots flying the Hawker Tempest and, later, the Gloster Meteor—the Royal Air Force's first operational jet fighter—were tasked with intercepting the missiles mid-flight. Some pilots, unwilling to risk gunfire in densely populated areas, resorted to flying alongside the buzz bombs and using their own wingtips to "tip" the V-1s off balance, sending them crashing into uninhabited fields or the English Channel. It was a desperate and dangerous maneuver, but one that underscored the severity of the threat and the ingenuity of those defending the homeland.

Over time, improved radar detection, enhanced anti-aircraft coordination, and refined interception tactics reduced the effectiveness of the V-1 campaign. But the damage had already been done. The V-1 flying bomb demonstrated that a pilotless, mass-produced aircraft could function not only as a tactical weapon but as a tool of psychological warfare. It foreshadowed a future in which missiles and drones could bypass traditional defenses and strike at the heart of civilian populations.

World War II was the crucible in which uncrewed aerial vehicles transitioned from speculative engineering projects to practical wartime systems. The Radioplane OQ-2 proved that drones could be standardized, built cheaply, and deployed in large quantities to meet training and tactical needs. The V-1, in contrast, showed that pilotless aircraft could be turned into deadly operational weapons—devices that could inflict significant damage without putting a single pilot at risk. Together, these two technologies marked the birth of modern unmanned aerial warfare.

The war catalyzed the drone field across every dimension—engineering, doctrine, logistics, and battlefield integration. And while both Allied and Axis systems still struggled with limitations in accuracy, control, and reliability, the strategic implications were clear. The genie had been released. Uncrewed flight had found its place in warfare, and it would not return to the realm of theory again.

As World War II progressed and Allied forces closed in on the last defensive bastions of Nazi Germany, military planners began to entertain increasingly unconventional strategies. Among the most daring—and ultimately tragic—was one of the earliest large-scale attempts to weaponize full-sized, radio-controlled aircraft for long-range precision strikes. Known as Project Aphrodite, this effort marked a grim chapter in the evolution of uncrewed aviation, revealing both the strategic desperation of late-war planning and the formidable technical hurdles that confronted early drone developers.

In contrast to the compact V-1 buzz bomb or the nimble Radioplane OQ-2 target drone, Project Aphrodite involved the conversion of full-sized, manned bombers into remotely piloted flying bombs. These were not small test aircraft but repurposed B-17 Flying Fortresses and B-24 Liberators—once the pride of the Allied strategic bombing campaign—now loaded with vast quantities of explosives and redirected toward hardened German targets. The aim was simple in concept but radical in execution: deliver an overwhelming payload against enemy fortifications without sacrificing a bomber crew in the process.

By mid-1944, conventional air strikes were proving ineffective against a range of critical Axis sites. Submarine pens along the French coast, V-1 and V-2 rocket launch installations, radar stations, and reinforced command bunkers had all withstood repeated bombardments. These heavily fortified positions absorbed attack after attack while continuing to pose serious threats to Allied operations. Something more precise—and more destructive—was needed.

Thus was born Operation Aphrodite, under the direction of the U.S. Army Air Forces, and its naval counterpart, Operation Anvil, managed by the U.S. Navy. The plan called for aging bomber airframes to be stripped of all unnecessary weight: armor, defensive guns, and crew stations were removed to accommodate up to 20,000 pounds of high explosives, usually modified Torpex charges. In their place, engineers installed RCA and Sperry radio control systems, gyrostabilized autopilot units, and television cameras in the nose—primitive video feeds that would allow remote pilots in trailing aircraft to visually steer the bombers toward their targets.

Each mission began with a crew of two men, who would take off and guide the flying bomb through its initial climb. Once over the English Channel, the crew would arm the warhead, activate the remote control systems, and parachute out—ideally over friendly territory or open water. From that point on, control shifted to a "mother" aircraft, often another B-17 or a Navy PV-1 Ventura, which would attempt to guide the bomber-drone the rest of the way to its target.

In theory, Aphrodite promised a way to inflict maximum damage on hardened enemy positions while minimizing Allied casualties. In practice, the missions were plagued by mechanical failures, navigation difficulties, and deadly accidents. Autopilot systems frequently malfunctioned, causing the drones to veer off course or respond unpredictably to commands. Radio signal interference and lag further hampered guidance efforts, making mid-flight course corrections nearly impossible. In several tragic cases, the explosive-laden bombers detonated prematurely—either due to electrical faults, structural stress, or human error—killing their onboard crews before they could bail out.

Despite the courage of the men involved and the ingenuity behind the technology, Project Aphrodite never achieved its intended operational success. Yet its legacy is important. It represented one of the first serious attempts to combine explosive payloads, remote navigation, and real-time video feedback in a single uncrewed platform. While the missions ended in failure more often than success, the lessons learned—about control systems, payload integration, and the psychological complexities of remotely operated combat—would echo in drone programs for decades to come.

One of the most tragic and high-profile episodes of Project Aphrodite involved Lieutenant Joseph P. Kennedy Jr., the eldest son of former U.S. Ambassador Joseph Kennedy Sr. and older brother of future President John F. Kennedy. On August 12, 1944, Kennedy and his co-pilot, Lieutenant Wilford Willy, were flying a modified B-24 Liberator as part of an Operation Anvil mission when disaster struck. Shortly after arming the aircraft's payload and before the bailout sequence could be executed, the bomber exploded in midair over the English countryside near Blythburgh, Suffolk. Both men were killed instantly. Investigators later concluded that a faulty arming circuit was the most probable cause of the premature detonation.

The death of Joseph Kennedy Jr., a decorated naval aviator and prominent member of a politically influential family, cast a long shadow over the Aphrodite project. His loss brought renewed public and military scrutiny to the viability and safety of the missions, and underscored the extreme risk inherent in combining experimental technology with live explosive payloads and human crews expected to parachute clear in time.

Between August 1944 and January 1945, more than a dozen Aphrodite and Anvil missions were launched. Almost none achieved their intended outcome. Drones veered off course, lost contact with control aircraft, or detonated prematurely before reaching their targets. Some aircraft crashed in rural or coastal areas without inflicting harm. Others came dangerously close to Allied positions, raising fears about collateral damage and the accidental loss of friendly lives. Despite the ingenuity behind the concept and the bravery of those involved, the program failed to produce any confirmed strategic success.

By early 1945, the project was quietly shut down. Conventional bombing raids, combined with the accelerating collapse of German infrastructure, made the high-risk drone missions increasingly unnecessary. But even in failure, Aphrodite marked a significant moment in the evolution of uncrewed aerial warfare.

It was the first known attempt to use live television video feeds to guide an aerial weapon in real time. It marked the first practical integration of airborne radio control systems in full-sized manned aircraft repurposed for uncrewed missions. Most significantly, it was the first operational deployment of drone-like aircraft intended to strike strategic targets deep behind enemy lines. These were bold, unprecedented steps—ones that pushed the limits of wartime technology and introduced new questions about ethics, feasibility, and control.

Yet the project also exposed critical shortcomings. World War II-era analog systems struggled to provide the precision needed for real-time remote control. The massive payloads, composed of volatile high explosives, lacked the fail-safes necessary to prevent accidental detonation. And the hybrid manned-unmanned structure of the missions—requiring live crews to fly partway before bailing out—blurred operational lines and introduced unacceptable risks.

Though Aphrodite was ultimately shelved, its legacy lived on. The hard-won lessons from those missions informed the next generation of uncrewed systems that would emerge during the Cold War—especially in the design of decoy drones, cruise missiles, and pilotless bombers intended for nuclear delivery. In retrospect, Project Aphrodite was not just a failed operation. It was a first step into a new age of warfare, one where machines would increasingly assume the burdens—and the dangers—once carried by human pilots.

Aphrodite offered more than a desperate wartime experiment—it offered a glimpse into a future where machines, not men, would be sent into the most dangerous corners of the sky. While its missions often ended in failure, the project seeded a vision of uncrewed warfare that would, in time, become reality. But that future would not arrive overnight. It would require decades of technical refinement, doctrinal shifts, and persistent visionaries willing to challenge the assumptions of traditional aviation.

Among those visionaries, few were more unlikely or more influential than Reginald Denny. A British-born World War I pilot who had found a second career as a Hollywood actor, Denny brought a showman's flair and an engineer's obsession to the world of remote-controlled flight. In the 1930s, he opened Reginald Denny Industries, a hobbyist shop on Hollywood Boulevard where model aircraft enthusiasts could build and fly their creations. But behind the storefront, Denny was conducting more than just civilian tinkering. He was transforming toy airplanes into prototypes for a new kind of military machine.

Denny began adapting his radio-controlled models for target practice, convincing the U.S. military that inexpensive drones could be used to train anti-aircraft gunners safely and effectively. His persistence led to the founding of the Radioplane Company, which produced the OQ-2—the first mass-manufactured drone in American history. Used extensively during World War II, the OQ-2 proved that unmanned aviation could be industrialized, standardized, and scaled for real-world utility. Denny's work bridged the gap between hobby and hardware, demonstrating that civilian innovation could serve the battlefield.

Another key figure in the early development of unmanned flight was engineer and inventor Elmer Ambrose Sperry. Known for his work in gyroscopic technology, Sperry played a foundational role in stabilizing both manned and unmanned aircraft. His mechanical gyroscopes, originally developed for naval gun targeting and ship navigation, soon found applications in aerial platforms—allowing aircraft to maintain directional stability and altitude without constant manual input. Sperry's systems were essential to the Kettering Bug, the earliest American cruise missile prototype, and later contributed to the autopilot capabilities of mid-century drones and flying bombs.

Sperry's legacy extended beyond basic stabilization. His advancements in inertial guidance laid the groundwork for modern UAVs capable of navigating without external signals. In today's uncrewed aircraft, inertial navigation systems remain critical for missions in GPS-denied environments, enabling autonomous operation even in the most hostile or remote areas of the globe. What began as a gyroscope in a torpedo became the beating heart of digital navigation.

No discussion of early drone history is complete without Charles Franklin Kettering, the prolific inventor and head of research at General Motors. A man of boundless curiosity and organizational skill, Kettering brought industrial rigor to military aviation at a time when innovation often relied on ad hoc experimentation. His Kettering Bug—a pilotless, gyroscopically stabilized biplane capable of delivering a high-explosive warhead over enemy territory—was the closest thing World War I ever saw to a cruise missile. Though it never reached combat before the Armistice, the Bug was decades ahead of its time.

Working alongside Sperry, Orville Wright, and later Carl Norden, Kettering assembled a multidisciplinary team that would prototype many of the control and guidance systems that future generations of UAVs would depend on. Their collaboration fused aviation with automation, creating the earliest blueprint for guided aerial weapons. While their machines remained rudimentary by modern standards, the ideas they embodied—stabilized flight, remote guidance, automated targeting—would ripple forward into the Cold War and beyond.

In these men—Denny the dreamer, Sperry the technician, and Kettering the industrialist—uncrewed aviation found its first champions. Together, they helped move the drone from the margins of military thought to the center of modern warfare's technological evolution.

Born in Baghdad in 1937 and raised in Israel, Abraham Karem served as an aeronautical engineer in the Israeli Air Force before emigrating to the United States in the 1970s. By the early 1980s, working out of his garage in California, Karem developed a UAV named Amber, a lightweight, long-endurance aircraft designed for persistent surveillance. Amber would eventually evolve into the MQ-1 Predator, a system that would transform not only aerial reconnaissance, but the very nature of modern warfare.

His company, Leading Systems Inc., was acquired by General Atomics after facing financial difficulties. Under the support of U.S. defense funding and intelligence interest, Karem's concepts were refined into a fleet of operational systems capable of loitering over targets for hours, streaming real-time video feeds, and later—delivering precision-guided munitions. The Predator was not just a drone; it was a paradigm shift. It allowed commanders to surveil and strike across the globe with minimal risk and unprecedented immediacy. In the conflicts of Iraq, Afghanistan, and elsewhere, it became the defining symbol of remote warfare—a technology that blurred the lines between pilot and operator, target and screen.

Though not a drone engineer by trade, Dr. John Paul Stapp, a U.S. Air Force physician and biophysicist, made an indirect yet profound contribution to unmanned aviation. Stapp became famous for his rocket sled tests at Holloman Air Force Base, where he subjected himself to forces exceeding 40 Gs in order to study the limits of human endurance in high-speed flight. His research helped shape safety standards for pilots and astronauts, but it also underscored the stark truth that certain flight regimes—especially those involving extreme acceleration, altitude, and velocity—were better suited to machines than men. In this way, Stapp's work added scientific justification to the growing argument for pilotless systems in aerospace applications.

At the policy and research level, John Stuart Foster Jr. played a pivotal role in enabling the drone revolution from behind the scenes. As Director of Defense Research and Engineering at the Pentagon and later in his leadership roles at DARPA and Lawrence Livermore National Laboratory, Foster championed the miniaturization of avionics and the integration of emerging sensor technologies. During the 1970s and 1980s, his guidance and funding priorities advanced key breakthroughs in micro-UAV development, encrypted data links, and autonomous navigation systems. His advocacy for forward-leaning military technologies created the conditions under which UAV platforms could flourish in the decades that followed.

Together, these pioneers laid the foundation for a new era of aerial capability. Through invention, sacrifice, and strategic foresight, they transformed uncrewed flight from a marginal pursuit into a core element of military power. The systems they inspired could persist over enemy territory, act without fatigue, strike without warning, and collect intelligence with a level of precision previously unimaginable. In time, their principles would be scaled, weaponized, and integrated not only into combat and surveillance but into civilian life itself—redefining the role of the aircraft in the modern world.

While the United States and the Soviet Union dominated the Cold War's global arms race, one smaller nation emerged as a quiet pioneer in the field of unmanned aerial vehicles. Confronted by immediate threats and operating under the constant pressure of limited manpower, Israel began developing drones not as theoretical war machines but as battlefield necessities. In the 1970s and 1980s, the Israeli approach to UAV development would not only reshape its own defense doctrine but profoundly influence the evolution of modern aerial warfare around the world.

The roots of Israel's UAV revolution lay in hard lessons learned during the Yom Kippur War of 1973. That October, Israel was caught off guard by a coordinated surprise attack from Egypt and Syria, forcing the Israel Defense Forces into a defensive posture under conditions of severe informational disadvantage. Despite eventual battlefield success, the war revealed a glaring vulnerability: the IDF's inability to gain rapid, accurate intelligence on enemy movements in real time. To overcome this, Israeli defense strategists turned to unmanned flight—not as a supplement to conventional reconnaissance, but as a new core capability.

In the wake of the war, a number of Israeli defense firms—including Tadiran and Israel Aircraft Industries (IAI)—were tasked with creating compact, tactical UAVs capable of operating from forward positions, transmitting live battlefield imagery, and guiding artillery or air strikes with greater accuracy. These efforts soon yielded two groundbreaking platforms that would cement Israel's status as a drone innovator.

The Tadiran Mastiff was among the first. A short-range UAV equipped with day and night cameras, including early infrared sensors, the Mastiff could relay live video back to ground commanders from beyond the front lines. Its impact on battlefield decision-making was immediate. With real-time visibility of enemy armor and troop concentrations, Israeli units could maneuver more effectively, respond faster, and direct firepower with new precision. The Mastiff proved that situational awareness no longer had to depend on manned overflight or delayed reconnaissance reports.

More ambitious still was the IAI Scout, introduced in the early 1980s. Larger and more advanced than the Mastiff, the Scout boasted improved range, stability, and sensor integration. Its twin-boom airframe and pusher-propeller configuration became a visual signature of Israeli drone engineering. Rugged and versatile, the Scout was designed for rapid deployment in field conditions. It could be launched from a catapult and recovered by parachute or belly skid, allowing frontline units to operate UAV support without the need for runways or elaborate logistics.

What set both platforms apart was not just their technological novelty, but their immediate utility. Operating at low altitudes with relatively quiet propulsion systems, these drones provided a persistent presence above contested terrain. Their imagery enabled real-time artillery corrections, airstrike coordination, and rapid assessment of enemy responses—all without endangering human pilots. For a country surrounded by adversaries on multiple borders, these capabilities offered a critical force multiplier.

The success of Israel's UAV operations did not go unnoticed. American military observers were particularly impressed by the effectiveness of drones during the 1982 Lebanon War, where Israeli forces used the Mastiff and Scout in concert with electronic warfare to blind Syrian radar and neutralize surface-to-air missile sites. The performance of these drones—providing surveillance, targeting, and post-strike damage assessment—convinced the U.S. Department of Defense that it had underestimated the potential of UAVs. The result was a renewed American investment in drone technology, including partnerships with Israeli firms, technology transfers, and the launch of joint development programs that would shape platforms like the RQ-2 Pioneer and, eventually, the Predator.

Israel's drone innovations were born not in think tanks or procurement cycles, but in the crucible of battlefield necessity. By integrating UAVs directly into combat doctrine—rather than treating them as afterthoughts—Israel proved that drones could serve not just as eyes in the sky, but as vital instruments of real-time warfighting. In doing so, it helped usher in the age of the tactical drone—a development that would alter the trajectory of armed conflict for generations to come.

The turning point for Israel's unmanned aerial doctrine came during the 1982 Lebanon War, when the Israel Defense Forces launched a decisive air campaign against Syrian military positions in the Bekaa Valley. Syrian surface-to-air missile batteries had been dug in across the region, protected by dense radar coverage and layered defenses. In response, Israeli planners executed one of the most sophisticated air-ground operations of the era—an assault that seamlessly integrated manned aircraft, electronic warfare assets, and for the first time, uncrewed aerial vehicles acting as tactical combat enablers.

The operation, codenamed Mole Cricket 19, marked the first modern use of drones as active participants in a combined-arms strike. The Israeli Air Force deployed Scout and Mastiff UAVs to conduct a variety of missions that had, until then, required risky manned overflight. The drones detected and mapped the locations of Syrian SAM sites, simulated aircraft profiles to provoke radar activation, and relayed real-time targeting data to Israeli fighter jets and artillery units waiting to strike. As the drones hovered silently over enemy territory, Israeli pilots exploited the surveillance feeds to destroy all nineteen Syrian missile batteries. In a matter of hours, the Syrians had lost their entire air defense network. The IDF went on to down more than eighty Syrian aircraft in dogfights—without suffering a single air loss of its own.

For the first time in military history, UAVs had not only contributed to victory but had made it possible. They conducted reconnaissance, jammed enemy systems, provided live bomb damage assessments, and did so without endangering Israeli aircrews. The implications were immediate and far-reaching. Militaries around the world—particularly the United States—took notice.

At the time of the Lebanon War, American UAV programs had stagnated. Decades of bureaucratic inertia, budget constraints, and shifting defense priorities had sidelined drone development within the U.S. armed forces. But the success of Mole Cricket 19 changed that overnight. U.S. military observers, especially within the Navy and Marine Corps, were deeply impressed by what they had seen. Within two years, the United States entered into a cooperative development agreement with Israel, seeking to adapt and adopt the technologies that had so dramatically reshaped the battlefield.

The result was the RQ-2 Pioneer, a short-range surveillance drone based heavily on the Israeli Scout platform. Built through a partnership between Israel Aircraft Industries and American firm AAI Corporation, the Pioneer became the primary UAV system for the U.S. Navy and Marine Corps in the late 1980s. Like its predecessor, it was catapult-launched, recovered via net capture aboard ships, and provided real-time imagery for surveillance, targeting, and battle damage assessment. The Pioneer gained particular fame during the Gulf War in 1991, when Iraqi troops reportedly surrendered to the circling drone—waving white flags at the small aircraft overhead, unaware that it was unmanned.

Israel's early drone systems set standards that few others could match. Their platforms delivered real-time video to ground stations, survived battlefield conditions, required compact logistical footprints, and could be deployed and recovered quickly from austere environments. But Israel's true innovation was not just technical. It was doctrinal. Israeli commanders embedded drones into frontline combat operations—not as auxiliary tools, but as core assets.

They pioneered the practice of assigning UAVs directly to tactical units rather than strategic headquarters. They used drones as decoys to draw out enemy air defenses, as jammers to disrupt radar and communications, and as persistent eyes in the sky for intelligence, surveillance, and reconnaissance. Their integration of UAV feeds with manned aircraft and ground units created a multi-layered ISR architecture that became a blueprint for drone-centric warfare.

Beyond the battlefield, Israel became a global powerhouse in drone production and export. Companies such as Elbit Systems, Israel Aerospace

Industries, and Aeronautics Ltd. began supplying UAVs to over fifty countries worldwide. Their offerings spanned a wide range—from small tactical drones like the Skylark to MALE-class systems such as the Heron, capable of long-endurance missions at medium altitudes. These exports often came with doctrine development packages, training, and maintenance support, allowing customer nations to integrate drones into their forces more effectively.

Perhaps Israel's most enduring contribution to drone warfare was conceptual. It redefined what a drone could be—not merely a support platform, but a decisive battlefield instrument. In doing so, it helped inaugurate a new era in military strategy, one in which uncrewed systems would no longer play a supporting role behind the scenes, but a leading role at the point of contact.

By the close of the twentieth century, drones had evolved far beyond their early reputation as experimental novelties or supplemental battlefield tools. No longer confined to the shadowy corners of intelligence agencies or the limited budgets of niche military programs, unmanned aerial vehicles had become an integral pillar of modern warfare. They enabled persistent surveillance, remote precision strikes, and an unprecedented degree of strategic autonomy. What began as little more than glorified radio-controlled aircraft had matured into a sophisticated class of machines that could observe, assess, and engage—all while being piloted remotely from control centers located continents away.

This transformation was made possible by a convergence of technologies that matured in the late 20th century. Global Positioning System (GPS) navigation allowed for pinpoint accuracy across vast distances. Miniaturized sensors and high-resolution optics gave UAVs the power to see with clarity once limited to manned reconnaissance aircraft. Digital telemetry enabled real-time communication between aircraft and command centers, while satellite links extended the reach of remote operations across hemispheres. Together, these breakthroughs allowed drones not just to fly, but to think and act within integrated battle networks. By the 1990s, what had once belonged to prototype programs and black-budget testing had entered the operational mainstream.

The roots of this revolution, however, stretched much further back. In the years following World War II, drone development continued—but slowly, and largely within military testing communities. Most postwar UAVs were still considered remote-controlled aircraft, lacking the autonomy or systems integration needed for frontline combat. They served primarily as target drones or experimental reconnaissance platforms.

Among the earliest postwar examples was the American JB-4, introduced in the 1940s. It combined a basic television camera and radio-command guidance system, making it one of the first real attempts at a guided strike drone. Though primitive by today's standards, it hinted at a future where visual targeting and remote engagement might one day converge. In Australia, engineers developed the GAF Jindivik, a high-speed aerial target drone that would serve in Commonwealth militaries for decades. Its successful performance helped validate the idea that jet-powered UAVs could be reliable, reusable, and relevant to modern air defense.

In the United States, progress continued in parallel. In 1951, the Teledyne Ryan Firebee I took to the skies, becoming one of the first jet-powered UAVs built in large numbers. Designed for both surveillance and target drone roles, the Firebee became a mainstay in U.S. aerial testing for decades. A few years later, Beechcraft entered the conversation with the Model 1001, a prototype UAV pitched to the U.S. Navy in 1955. While that specific model never reached production, it reflected growing interest from civilian aerospace firms in entering the unmanned systems space.

Despite incremental technical advancements, UAVs during the 1950s remained largely within the realm of sophisticated remote-control aircraft. But global politics soon forced evolution. In 1959, after a string of manned reconnaissance flights over hostile territory resulted in increased pilot losses, the U.S. Air Force began formal planning for high-risk missions that would remove humans from the cockpit. That need became urgent on May 1, 1960, when the Soviet Union shot down a CIA-operated U-2 spy plane flown by Francis Gary Powers. The international fallout was immediate and severe. The Cold War had entered a new phase, and the vulnerabilities of manned overflight had been laid bare.

Within days of the U-2 incident, the U.S. Air Force launched a classified program under the codename Project Red Wagon—an effort to accelerate

UAV development for strategic reconnaissance missions. It marked a turning point in how drones were viewed by the military establishment. What had been experimental or peripheral was now considered essential to national security.

The Vietnam War further accelerated the transition of UAVs from research platforms to combat-tested assets. Following the Gulf of Tonkin incident in 1964, U.S. military operations escalated sharply, and UAVs were deployed in increasingly hazardous roles. Platforms like the Ryan Model 147 Lightning Bug were sent on high-risk reconnaissance missions over North Vietnam and along the Chinese border. The Ryan AQM-91 Firefly and the supersonic Lockheed D-21 drone pushed the boundaries of altitude, speed, and survivability. While many of these missions remained classified, their scope was vast. When UAVs were lost over hostile territory, official U.S. responses were deliberately vague—often a simple "no comment." Behind the scenes, however, the data they collected shaped targeting decisions, bombing campaigns, and the broader strategic picture of the war.

Though little public recognition existed at the time, the UAVs of the Vietnam era proved that drones could survive, operate, and contribute meaningfully to warfare in contested airspace. Their quiet success laid the groundwork for the major leap to come: the development of drones that were not only persistent eyes in the sky, but armed extensions of national policy.

During the 1967–1970 War of Attrition, Israeli intelligence services began quietly experimenting with lightweight, tactical unmanned aerial vehicles equipped with cameras. These early drones were designed to launch from short, improvised airstrips and carry out brief overflights across the Suez Canal into Egyptian-held territory. Though rudimentary in form, they returned valuable photographic intelligence—images of troop concentrations, artillery emplacements, and logistical movements that would have been otherwise unobtainable without risking manned reconnaissance flights.

The effectiveness of these uncrewed aircraft prompted further development, and by the time of the 1973 Yom Kippur War, Israel was using drones not only for reconnaissance but for active deception. In the opening stages of the conflict, Israeli forces launched UAVs as decoys, sending them into enemy airspace to trigger Egyptian and Syrian anti-aircraft systems. These drones, devoid of sensors or weapons, served a singular purpose: to provoke enemy radar operators into firing expensive surface-to-air missiles at worthless targets. It was a brilliant tactical adaptation that helped reduce the threat posed to Israeli pilots during follow-up air operations.

Following the war, several of the engineers who had worked on these drone projects left the military to form a private venture dedicated to unmanned systems development. The company was eventually acquired by Tadiran, a major player in Israeli defense electronics, giving birth to the country's first domestically produced UAV programs. These early collaborations between private industry and military innovators laid the foundation for what would become one of the most advanced drone industries in the world.

Meanwhile, on the other side of the globe, the United States was also deepening its reliance on UAVs—albeit in a more secretive fashion. By 1973, the U.S. government officially acknowledged the use of drones in Southeast Asia, particularly in Vietnam, Laos, and Cambodia. Most missions focused on reconnaissance, damage assessment, and electronic surveillance. The scale of the effort was significant. The 100th Strategic Reconnaissance Wing alone conducted over 3,400 UAV sorties, a number that underscored the growing demand for risk-free intelligence collection in hostile territory.

The risks, however, remained considerable. By the end of the Vietnam conflict, 554 UAVs had been lost to enemy action, mechanical failure, or retrieval issues. But those losses paled in comparison to the toll among human aircrews: more than 5,000 American airmen were killed during the war, with over 1,000 captured or declared missing. These figures fueled a growing consensus within the military: drones could reduce casualties—and were increasingly worth the investment.

Key figures in the Air Force voiced their support. In 1972, General George S. Brown of the USAF Systems Command declared, "The only reason we need UAVs is that we don't want to needlessly expend the man in the cockpit." That same year, General John C. Meyer of Strategic Air Command stated bluntly, "We let the drone do the high-risk flying. The loss rate is high, but we're willing to risk more of them—they save lives." Such remarks reflected a doctrinal shift that had been slow to take hold, but was now gaining traction at the highest levels of command.

In Israel, UAV development also took on new urgency after the 1973 war. Having suffered heavy aircraft losses to Soviet-supplied surface-to-air missile systems, the Israeli Air Force accelerated efforts to find safer alternatives for real-time battlefield surveillance. This renewed push culminated in the development of the IAI Scout—the first Israeli drone to offer real-time video transmission to ground commanders. Introduced in the early 1980s, the Scout would become a central asset in Israeli tactical operations and later serve as the basis for international collaboration with the United States.

The era of drones as battlefield supplements was ending. They were becoming integral to operational planning, capable of shaping outcomes in both air and ground campaigns. What had begun as reconnaissance curiosities were evolving into indispensable components of modern warfighting.

During the 1982 Lebanon War, Israel's UAVs proved not only useful but decisive. Scout and Mastiff drones operated in tandem with manned aircraft and electronic warfare platforms to deliver a comprehensive air campaign that left Syrian air defenses in ruins. The drones provided real-time imaging to Israeli commanders, allowing them to assess the battlefield moment by moment. They deployed radar decoys to mislead Syrian surface-to-air missile crews, drawing fire away from piloted aircraft. Most critically, they helped orchestrate the complete neutralization of Syria's air defense network without the loss of a single Israeli pilot. It was a watershed moment in drone warfare—an integrated operation in which UAVs shifted from support roles to indispensable combat enablers.

Even as battlefield drones matured operationally, Israeli engineers were already looking ahead. By 1987, experimental UAV platforms were demonstrating flight characteristics that rivaled manned fighter jets. These next-generation testbeds explored post-stall maneuvering, three-dimensional thrust vectoring, and radar-deflecting airframe geometries. Long before stealth drones became a headline, Israeli researchers were engineering airframes that reduced radar signatures and performed agile, evasive maneuvers—all without a human pilot on board. The future of air combat was quietly taking shape.

Meanwhile, in the United States, a strategic transformation was unfolding. With the Cold War winding down and a new era of regional conflicts emerging, U.S. defense planners began reassessing their force posture. The conflicts of the 1990s—in the Balkans, the Persian Gulf, and East Africa—exposed a common need: real-time intelligence, low-risk surveillance over hostile terrain, and precise targeting for limited-scale engagements. Traditional platforms were expensive, slow to deploy, and often carried political risks. The solution would come not from a new fighter jet or spy satellite, but from a drone.

The breakthrough arrived with the MQ-1 Predator. Developed by General Atomics Aeronautical Systems under the guidance of Abraham Karem—whose early work on the Israeli Amber drone had laid the foundation—the Predator fused long-endurance flight, satellite uplinks, and real-time imaging into a single, modular platform. What began as a reconnaissance drone would soon become a weapon of precision and persistence, capable of executing the modern military mantra: "See everything. Strike selectively. Risk nothing."

The Predator took its first flight in 1994 and was deployed over Bosnia the following year. It offered a capability that traditional reconnaissance aircraft could not: continuous surveillance for more than twenty-four hours at a time, without placing a pilot in harm's way. Equipped with electro-optical and infrared cameras, the drone relayed live feeds to operators across the globe, providing commanders with round-the-clock situational awareness in environments where U.S. aircraft had limited access. Though initially unarmed, the Predator's utility in intelligence gathering and targeting made it a staple in operations across the Balkans.

By 2001, the Predator had evolved from observer to hunter. Fitted with AGM-114 Hellfire missiles, it became the first armed drone system in the U.S. inventory. Its operational debut came during the early stages of Operation Enduring Freedom, when a Predator launched a missile strike against a high-value al-Qaeda target in Afghanistan. The event marked the beginning of a new era in warfare—one in which the kill chain could be completed without ever leaving the ground.

The Central Intelligence Agency, already deeply familiar with drone operations from its earlier work with the D-21 and AQM-34 Firebee programs, quickly incorporated armed UAVs into its counterterrorism doctrine. Flying from clandestine airfields and operated by civilian contractors under classified protocols, CIA drones became central to the agency's targeted killing strategy. In the aftermath of 9/11, they emerged as the tool of choice for locating, monitoring, and eliminating terrorist leaders across the Middle East, South Asia, and North Africa.

Under the Joint Special Operations Command, these operations were further refined into "find, fix, finish" missions. Drones would locate a target, confirm identity and behavior, and strike with lethal precision—often within minutes of receiving final approval. Operators sat in remote ground control stations, sometimes located as far away as Creech Air Force Base in Nevada, piloting aircraft that hovered silently over warzones thousands of miles away. Satellite constellations relayed real-time video and control signals, creating a global network through which surveillance and warfare could be conducted with unprecedented reach.

The 1980s and 1990s saw a revolution not only in drone doctrine, but in the technologies that made such operations possible. Advances in microelectronics, composite materials, data transmission, and sensor integration enabled drones to do more with less—to fly farther, see clearer, and act faster than ever before. These capabilities captured the full attention of both the U.S. military and intelligence communities. Drones were no longer experimental. They had become vital instruments in the new age of precision, asymmetrical warfare.

Within the Central Intelligence Agency, a quiet but consequential shift was taking place. The agency's Counterterrorism Center (CTC), already familiar with the limitations of traditional intelligence-gathering tools, began exploring the use of unmanned aerial vehicles for high-risk, precision operations. UAVs offered a unique combination of persistence, discretion, and flexibility—qualities that suited the agency's evolving mission in the post-Cold War landscape. No longer confined to reconnaissance alone, drones were being envisioned as surgical instruments of national policy.

At the same time, the U.S. Department of Defense launched its own initiatives to revitalize tactical UAV development. One key effort emerged through a partnership between AAI Corporation, a U.S. aerospace firm, and Malat, the UAV division of Israel Aircraft Industries. Drawing upon Israeli operational experience and American industrial capacity, this collaboration produced the AAI Pioneer—a tactical surveillance drone designed to support real-time battlefield awareness.

The Pioneer made its combat debut during the 1991 Gulf War. Catapult-launched and net-recovered from Navy ships and forward operating bases, the drone provided live video intelligence to U.S. and coalition forces, enhancing targeting accuracy and force protection without putting pilots at risk. In one now-famous episode, Iraqi soldiers reportedly surrendered to a circling Pioneer, mistaking it for a precursor to an airstrike. The incident underscored not only the psychological impact of drones, but their emerging place in modern warfare. The Pioneer's success cemented the idea that UAVs were not experimental anymore—they were operational assets.

While early drone generations focused on surveillance, the late 1990s marked a decisive transition toward armed UAVs. The General Atomics MQ-1 Predator became the centerpiece of this evolution. Built upon the foundational work of Abraham Karem and his Amber drone, the Predator incorporated long-endurance flight, satellite data links, and real-time sensor feeds. When fitted with AGM-114 Hellfire missiles, the Predator completed its transformation—from passive observer to active combatant. It was no longer just an eye in the sky, but a remote-controlled hunter with lethal reach.

The Predator's weaponization redefined the conduct of air warfare and counterterrorism strategy. In Afghanistan, Iraq, Yemen, and Somalia, the drone became a central instrument of U.S. foreign policy, conducting targeted strikes against high-value individuals and insurgent networks. Its ability to loiter undetected, gather intelligence, and execute pinpoint strikes made it invaluable in environments where traditional airpower or ground raids carried higher risks.

Recognizing the growing strategic utility of drones, other nations and alliances began ramping up their own UAV development. Between 2002 and 2005, the European Union funded CAPECON, a multi-nation research initiative aimed at advancing UAV design, deployment protocols, and airspace integration across Europe. Although smaller in scale than U.S. or Israeli programs, CAPECON reflected a broader recognition that drone technology was becoming essential—not only for military operations, but for border security, disaster response, and environmental monitoring.

By 2012, the United States Air Force operated 7,494 UAVs—nearly one-third of its total aircraft inventory. That figure did not include the CIA's own fleet, which operated in parallel for clandestine missions in Pakistan, Yemen, and the Horn of Africa. While military drones focused on battlefield ISR and precision strike, CIA operations emphasized covert surveillance and targeted killings under Title 50 authority. Often flown from undisclosed locations by civilian contractors, these missions became a defining feature of the post-9/11 counterterrorism landscape.

Drone technology, once a closely held specialty of just a few nations, had begun to spread. By 2013, more than fifty countries had developed or deployed unmanned aerial vehicles. Among the most prominent were China, Iran, Israel, Pakistan, Turkey, and Russia—nations that had either reverse-engineered foreign systems or built domestic programs from the ground up. As UAV proliferation accelerated, it became increasingly difficult for global watchdogs, intelligence agencies, or arms control groups to track every drone system in active use. What had begun as a novel tool of asymmetric advantage had become a near-universal capability—one whose implications would ripple far beyond the battlefield.

Among the most ambitious ventures in the evolution of miniature uncrewed systems is the Micromechanical Flying Insect, or MFI—a palm-sized robotic fly designed to perform covert surveillance and reconnaissance missions in places where traditional UAVs cannot operate. Initiated in 1998 by a research team at the University of California, Berkeley, the MFI project received early and sustained support from the U.S. Office of Naval Research and the Defense Advanced Research Projects Agency (DARPA). Its strategic promise was immediately evident: a biologically inspired, autonomous platform capable of penetrating spaces too dangerous, too cramped, or too sensitive for manned aircraft or even conventional drones.

Unlike larger UAVs designed for endurance, payload delivery, or altitude dominance, the MFI was envisioned for an entirely different role. Its mission was not to fly high or carry weapons, but to blend in—serving as a near-invisible reconnaissance tool in tight quarters, urban interiors, or hostile territory. Its potential made it particularly appealing to military and intelligence agencies who imagined scenarios in which an insect-sized drone could act as a literal "fly on the wall"—providing live video or environmental data from enemy compounds, terror cells, or battlefield ruins without alerting those nearby. In this vision of future surveillance, size was not a limitation, but a decisive advantage.

The MFI's possible uses extended beyond warfare. Scientists imagined microdrones like it being deployed in collapsed buildings after earthquakes, exploring nuclear disaster zones, or even mapping the surfaces of other planets. Their compact scale and low logistical demands made them ideal for missions requiring stealth, precision, and endurance under extreme conditions.

Early prototypes of the MFI were remarkably small. Weighing just 100 milligrams and spanning roughly two centimeters from wingtip to wingtip, the initial designs used stainless steel beams and polymer joints to replicate the mechanical structure of an insect thorax. However, these materials quickly proved too heavy, limiting the drone's lift capacity and making sustained flight impossible. Researchers responded by switching to ultra-light carbon fiber beams and micromachined silicon joints—materials that not only reduced weight but allowed for more precise, biologically accurate movement. The wings, designed to flap in coordinated rhythm like those of a true insect, were connected to electrostatic actuators that mimicked the micro-muscle function of a fly.

What made the MFI especially promising was its extraordinary cost-efficiency. Though advanced in concept and fabrication, the raw materials for each unit were estimated at only ten cents. This made the platform inherently scalable, suitable for swarm-based operations in which dozens or hundreds of units could be deployed simultaneously. In scenarios where some units might fail or be destroyed, the cost per asset was negligible—an important consideration for missions involving high risk or expendability.

Because of its size, the MFI could not be controlled by traditional radio or satellite links. Instead, it was designed to operate fully autonomously, relying on a compact suite of onboard systems to interpret its environment and respond in real time. The control architecture was as sophisticated as it was miniaturized. Solar-rechargeable battery packs powered the insect's systems, while two front-mounted optical sensors functioned as simple "eyes," feeding object-detection algorithms to help navigate complex surroundings. Environmental sensors monitored temperature, wind, and velocity, while onboard actuators controlled each wing in precise opposition or harmony, depending on the flight maneuver required. At the heart of it all, a signal processor converted navigational commands into wingbeat instructions, translating three-dimensional thrust vectors into calibrated aerodynamic responses.

To execute even a seemingly simple command like "fly forward," the MFI had to analyze its position in space, determine its desired trajectory, and dynamically adjust wingbeat asymmetry to maintain both lift and heading. All of this occurred in real time, without external direction or pilot input. In essence, each MFI functioned as a self-aware insect—engineered not merely to fly, but to survive, maneuver, and relay data from places no other machine could reach.

The MFI stood as a testament to the convergence of robotics, biology, and military necessity. In a world where information was often more valuable than firepower, and where access was everything, the ability to insert a surveillance asset no larger than a housefly into a hostile room represented a revolution. It offered a glimpse of future warfare—and future exploration—where the smallest machines might carry the greatest strategic weight.

Creating robotic insects at this scale has proven to be one of the most technically demanding frontiers in aerospace engineering. While the concept of the Micromechanical Flying Insect remains elegant in theory, its execution has revealed a host of complex challenges. Among the most persistent is the problem of processing visual data. The MFI's primitive optical systems—essentially low-resolution micro-cameras—often produce noisy, inconsistent input. For a machine that must make real-time navigational decisions based on visual cues, this kind of signal interference can lead to erratic or unstable flight behaviors. Developers have spent years refining algorithms capable of filtering this visual noise, stabilizing flight patterns, and enabling the drone to identify obstacles, recognize boundaries, and maintain orientation in cluttered environments.

Another formidable hurdle is hovering. While it may appear simple, hovering is one of the most biologically and mechanically demanding maneuvers in insect flight. Maintaining altitude in a fixed position requires a delicate balance of thrust, real-time corrections for wind and microturbulence, and near-instantaneous feedback from environmental sensors. Even slight lag in processing or actuator response can cause the MFI to drift, lose lift, or collide with surrounding objects. Achieving reliable, sustained hovering at insect scale remains an elusive goal.

Adding to the complexity is the challenge of replicating the nonlinear aerodynamics inherent in natural insect flight. Real flies exploit aerodynamic phenomena such as delayed stall, rotational lift, and wake capture—subtle forces that allow them to generate lift in ways that fixed-wing aircraft cannot. These effects are still the subject of active scientific study and are extraordinarily difficult to model in computational systems, let alone reproduce in real time through artificial actuators. Translating these behaviors into silicon-based flight control systems has become a cutting-edge effort in both robotics and biomechanics.

The MFI's development timeline reflects the incremental nature of this innovation. The program began in 1998 when UC Berkeley researchers received a $2.5 million DARPA contract to explore biologically inspired microflight. By 2001, the first prototype demonstrated thrust from a single wing mounted on a test stand—a small but critical step. In 2002, the design shifted from stainless steel components to ultra-light carbon fiber, dramatically improving the lift-to-weight ratio. A year later, a test rig generated 500 microNewtons of lift from a single flapping wing, enough to support initial test models. From 2004 onward, the focus shifted to reducing overall frame weight, improving actuator efficiency, and refining the autonomous algorithms needed for flight stability, navigation, and obstacle avoidance.

Despite the painstaking pace, the MFI remains at the vanguard of micro-UAV research. It represents a broader technological trend—the miniaturization of surveillance and intelligence platforms alongside the push for deeper operational discretion. In an era increasingly defined by urban warfare, irregular combat zones, and hard-to-access targets, the ability to deploy surveillance assets at insect scale could change the calculus of intelligence gathering.

Though still experimental, the breakthroughs achieved through the MFI program have already influenced how engineers think about flight at micro scale. Its integration of biological mechanics with robotic autonomy may one day yield a class of machines that operate with the stealth of a mosquito and the cognition of a drone. In the long arc of aerial innovation, the MFI stands as both a marvel of engineering and a portent of the future—a future where machines no larger than a thumbnail carry out missions once reserved for satellites, aircraft, or human agents.

As drone hardware became smaller, cheaper, and more accessible, civilian applications experienced an unprecedented boom. By the early 2010s, quadcopters and multirotor drones had become a global phenomenon, propelled by advances in battery technology, stabilization systems, and onboard cameras. Enthusiasts around the world embraced these compact flyers for a wide range of recreational purposes—first-person video (FPV) flight, aerial photography, landscape videography, and the adrenaline-fueled sport of drone racing. Once the preserve of military laboratories and defense contractors, uncrewed aerial systems had entered the mainstream.

The commercial and general aviation sectors, however, expanded more cautiously. Regulatory constraints, particularly those imposed by the Federal Aviation Administration (FAA) in the United States and the European Union Aviation Safety Agency (EASA), imposed strict limitations on drone operations. Most jurisdictions required visual line-of-sight operation and prohibited autonomous flights beyond the pilot's immediate control. These policies were intended to ensure safety in shared airspace, but they also hindered progress toward large-scale drone integration.

Technical limitations compounded the issue. Full autonomy—required for complex missions in dense urban areas or uncontrolled environments—remained elusive, as did the infrastructure for safely integrating drones into national air traffic control systems. Detect-and-avoid technology, reliable failsafe mechanisms, and real-time telemetry had to mature before commercial drone fleets could share skies with manned aircraft.

Nevertheless, innovation continued. Even under regulatory limitations, drones began appearing in agriculture, used to monitor crop health, spray fields, and optimize irrigation. In construction and surveying, drones created high-resolution 3D maps in a fraction of the time required by traditional methods. Emergency services used UAVs for search and rescue in disaster zones, accessing dangerous or obstructed terrain. Utility companies deployed them to inspect power lines, pipelines, and bridges—tasks previously requiring helicopters, climbers, or costly shutdowns. Logistics firms began testing drone delivery in rural areas, paving the way for autonomous package networks once regulatory environments caught up. Slowly but steadily, the groundwork for mass-scale commercial deployment was being laid.

Yet as civilian drone use soared, the darker implications of autonomous flight were beginning to materialize. In 2020, during the ongoing Libyan civil war, reports emerged of a Turkish-built Kargu-2 drone engaging and attacking a human target without direct human command. Designed by STM, a defense technology firm based in Ankara, the Kargu-2 is a loitering munition capable of autonomously identifying, tracking, and striking designated targets using onboard sensors and machine learning algorithms.

According to a United Nations Security Council report released in 2021, the drone may have executed a lethal strike against a retreating combatant, operating independently once activated. If confirmed, it would represent the first known instance of an AI-powered lethal autonomous weapon system (LAWS) acting without real-time human oversight. The incident immediately drew international concern and condemnation. The notion of drones making life-and-death decisions without human input raised alarm in military, legal, and human rights communities alike.

The term "killer robot," once the stuff of speculative fiction, now appeared in serious diplomatic briefings. Critics warned of a future where autonomous weapons could be mass-produced, deployed in swarms, and used without clear accountability. The Libyan incident underscored the growing urgency for global norms, treaties, and safeguards governing the use of artificial intelligence in lethal systems. Whether or not the Kargu-2 acted entirely on its own remains a matter of technical and legal debate, but the implications were undeniable. Autonomous drones had crossed a threshold—from tools of human intention to agents of autonomous force.

The emergence of autonomous lethal drones sparked an urgent and growing debate within the international community. Human rights organizations, legal scholars, military ethicists, and policymakers began sounding the alarm over what many considered a dangerous new arms race—one not just of firepower or range, but of decision-making itself. The core question was stark: should machines be allowed to make decisions over life and death?

At the heart of the issue was the concept of meaningful human control. Critics argued that delegating lethal authority to algorithms violated core principles of international humanitarian law, including distinction, proportionality, and accountability. Without a human in the loop, how could a drone properly evaluate whether a target posed an imminent threat? Who would be held responsible if an autonomous system committed a war crime? Would accountability lie with the manufacturer, the commander, or the machine?

The United Nations took notice. Beginning in the 2010s, a series of meetings under the Convention on Certain Conventional Weapons (CCW) convened state representatives, non-governmental organizations, and technical experts to discuss the implications of Lethal Autonomous Weapons Systems (LAWS). While consensus was slow to emerge, a growing number of member states—led by countries such as Austria, Brazil, and New Zealand—called for an outright preemptive ban on autonomous weapons before they could be widely deployed.

Meanwhile, a coalition of civil society organizations launched the Campaign to Stop Killer Robots, a global advocacy effort aimed at pushing governments toward a treaty prohibiting the development and use of fully autonomous weapons. The campaign gained endorsements from prominent figures in science, technology, and ethics, including Stephen Hawking and Elon Musk, who warned of the destabilizing effects such systems could have on global security.

Supporters of a ban cited numerous risks. Autonomous drones could be hacked, spoofed, or repurposed by bad actors. They might lower the threshold for entering armed conflict by reducing political risk to the aggressor. In the hands of authoritarian regimes, they could be used for domestic repression, extrajudicial killings, or population control. And in complex environments—such as crowded cities or refugee camps—they might struggle to distinguish combatants from civilians, making compliance with international law all but impossible.

Yet not all nations agreed. The United States, Russia, and China resisted efforts to establish a formal ban, arguing that existing legal frameworks already covered emerging weapons technologies. Proponents of autonomy within military planning cited potential benefits, such as faster response times, reduced casualties for friendly forces, and the ability to operate in communications-denied or high-threat environments. Some argued that properly designed autonomous systems could even reduce collateral damage by reacting faster and with greater precision than human operators.

This clash of perspectives created a legal and diplomatic impasse. While most agreed that some form of regulation was needed, few could agree on where to draw the line. Was a drone that independently tracked a target but required human approval to fire still considered autonomous? What about one that selected its targets from a pre-approved list? As AI technology advanced and human-machine collaboration became increasingly fluid, the boundary between assisted and autonomous warfare began to blur.

By the early 2020s, autonomous weapons had not yet been banned—but neither had they been meaningfully constrained. Development continued across multiple nations, and prototypes were tested in conflicts from the Middle East to North Africa. As drones became smaller, faster, and more intelligent, the urgency to establish ethical guardrails grew. But in a world of strategic competition and fragmented international consensus, progress remained uneven.

In the absence of a binding treaty, the future of autonomous warfare would be shaped not just by diplomacy, but by technology itself. And with each new advance in machine perception, target recognition, and decision-making algorithms, the window for regulation narrowed. The question was no longer whether killer robots were coming. It was whether humanity would agree on the rules before they arrived.

Among the most transformative developments in modern unmanned flight has been the widespread adoption of the quadcopter—a four-rotor aerial vehicle whose simplicity, agility, and reliability have made it the dominant form for small drones worldwide. Also known as a quadrotor or quadrocopter, the quadcopter design blends aerodynamic efficiency with mechanical minimalism, allowing vertical takeoff, hovering, and precision maneuvering without the complexity of traditional helicopters.

Though the idea of quadrotor flight dates back nearly a century, it remained largely experimental until the digital age. Early 20th-century engineers explored multi-rotor aircraft in their search for stable vertical lift, testing a variety of manned VTOL prototypes during the 1920s and 1930s. These designs showed promise—particularly in torque cancellation and rotor simplicity—but were plagued by significant drawbacks. Heavy and underpowered, with poor lift-to-weight ratios and no onboard stabilization systems, they demanded constant manual correction from their pilots. The workload was too great, the reliability too low, and the promise of the concept remained unrealized.

That began to change with the advent of compact, high-efficiency brushless motors, lightweight battery systems, and miniaturized electronics. The same digital revolution that powered smartphones and GPS navigation also made it possible to embed gyroscopes, accelerometers, and flight control algorithms into palm-sized drones. With the help of real-time electronic stabilization, the quadcopter's symmetrical rotor design finally came into its own—not as a manned aircraft, but as the ideal foundation for unmanned systems.

The engineering is elegant. Each of the quadcopter's four rotors generates lift and torque. Two spin clockwise, two counterclockwise. The opposing rotations cancel out the net torque that would otherwise cause the drone to spin uncontrollably—removing the need for a tail rotor and allowing for a more compact and balanced design. Flight is controlled not by adjusting blade pitch, as in traditional helicopters, but by varying the speed of each motor. Tilting forward or back—pitch—is achieved by increasing thrust on either the rear or front pair of rotors. Rolling left or right involves adjusting thrust between the side rotors. Rotating the aircraft—yaw—is handled by changing the torque balance between the clockwise and counterclockwise blades. This architecture reduces mechanical complexity and increases durability, while transferring the burden of stability and control to the flight controller's software.

The result is a highly maneuverable, modular platform capable of precise motion in confined spaces. From casual hobbyists to professional surveyors and military operators, users have embraced the quadcopter for its versatility. Its ability to hover and hold position makes it ideal for aerial photography and videography. Its small size allows it to fly beneath bridges or inside collapsed buildings for inspection and search-and-rescue operations. In military contexts, quadcopters have proven invaluable for close-range reconnaissance and tactical intelligence gathering, offering real-time situational awareness without the exposure of a human scout.

The commercial implications are equally profound. Companies have begun using quadcopters for infrastructure inspection, powerline monitoring, mapping, and even package delivery. Agricultural applications include crop assessment, precision spraying, and irrigation planning. In urban environments, quadcopters are being tested for traffic monitoring and air quality sensing. Their modular design allows for rapid customization—swapping out sensors, payloads, or cameras to suit mission-specific needs.

The rise of the quadcopter marks a convergence between early aviation imagination and 21st-century technological capability. Once dismissed as an impractical curiosity, the four-rotor aircraft has reemerged as the backbone of the global drone ecosystem. In doing so, it has democratized aerial access—putting the sky within reach of anyone with a few hundred dollars and a controller. More than just a clever piece of engineering, the quadcopter is a symbol of how modern innovation can breathe new life into old ideas, transforming them into tools of discovery, defense, commerce, and creativity.

Hexacopters, Octocopters, and Multirotor Variants

As drone technology has advanced, so too have the multirotor configurations designed to meet increasingly demanding missions. While the four-rotor quadcopter remains the most recognized and widely available format for recreational and light commercial use, larger multirotor systems—specifically hexacopters and octocopters—have emerged as essential tools in professional, industrial, and military domains. These platforms offer superior lift capacity, redundancy, and flight stability, expanding the reach and utility of unmanned aerial systems far beyond the capabilities of their smaller cousins.

Hexacopters, equipped with six rotors arranged in a circular pattern, occupy the middle ground between agility and muscle. Their primary advantage lies in their increased lift capacity, which enables them to carry heavier payloads such as high-resolution cinema cameras, lidar scanners, and multispectral imaging arrays. This makes them a preferred platform for aerial cinematographers, infrastructure inspectors, and agricultural analysts who rely on stable, high-fidelity data collection. The added rotors also provide greater resistance to wind and turbulence, making the aircraft more reliable in unpredictable conditions.

Perhaps most critically, hexacopters offer a level of redundancy that quads cannot match. In the event of a motor or propeller failure, a properly configured hexacopter can remain airborne long enough to return to base or execute a controlled landing. This built-in survivability is invaluable when operating over urban environments, critical infrastructure, or remote terrain where equipment loss or crash landings could be catastrophic. Platforms such as the DJI Matrice series have capitalized on this design, providing a stable, adaptable workhorse for high-end commercial and government users alike.

Beyond the hexacopter lies the octocopter—an eight-rotor powerhouse designed for maximum lift and mission durability. Octocopters are typically arranged in one of two configurations: a traditional eight-arm radial layout or an X8 configuration with four arms supporting two stacked motors apiece. These aircraft are designed not for speed or maneuverability, but for raw power, endurance, and safety. They can carry some of the heaviest payloads in the drone world, including RED and ARRI cinema cameras, emergency medical supplies, sensor arrays, or even classified military gear.

Their stability under load is unmatched. In high winds, high altitudes, or high-stakes operations, octocopters deliver smooth, predictable flight that is essential for precision tasks such as search and rescue, disaster assessment, or detailed 3D mapping. Many models are engineered with multiple fail-safes, allowing them to maintain flight even after losing one or two motors—an advantage that can make the difference between mission success and failure in critical deployments.

While octocopters are less portable and more power-hungry than smaller drones, their ability to execute complex missions with heavy payloads and high reliability makes them indispensable in certain industries. In fields where equipment safety, mission assurance, and operational flexibility are non-negotiable, octocopters have become the gold standard for unmanned aerial platforms.

Together, these multirotor systems illustrate the adaptability of drone architecture. From compact quadcopters to the heavy-lift octocopters used in military logistics and cinematic production, the drone landscape has diversified to meet the full spectrum of modern aerial needs. In doing so, it has redefined what is possible in the sky—scaling from the palm of a hobbyist's hand to the reach of industrial innovation.

Multirotor Trade-Offs

Each additional rotor in a multirotor system brings both advantages and trade-offs. Adding motors increases lift capacity, redundancy, and control precision, but it also adds weight, draws more power, and raises both cost and complexity. The result is a spectrum of drone configurations, each optimized for a specific balance of agility, endurance, and mission capability.

Quadcopters remain ideal for speed, portability, and general consumer use. Their minimalist design and relatively low energy demands make them the go-to choice for recreational pilots, aerial photographers, and lightweight commercial tasks. Compact and efficient, they can be launched quickly and maneuvered in tight quarters, making them suitable for environments where access and setup time are critical.

Hexacopters represent a middle ground—combining enhanced stability and payload capacity with a manageable footprint. They offer increased resistance to rotor failure and wind disturbance, along with sufficient power to carry professional-grade sensors or imaging equipment. For users who require more than a hobby-grade drone but do not need industrial-scale lift, the hexacopter offers a compelling balance of capability and practicality.

Octocopters, at the upper end of the multirotor spectrum, are engineered for demanding missions that require maximum redundancy and lift. They excel in carrying heavy loads, maintaining stability in rough conditions, and executing long-duration or high-stakes flights. However, they do so at the cost of power efficiency, portability, and system complexity. Deploying an octocopter often involves greater logistical preparation, including larger batteries, more frequent maintenance, and the need for skilled operators.

Choosing the appropriate multirotor configuration is ultimately a matter of mission parameters. Payload weight and type, desired flight time and range, environmental conditions, and tolerance for system failure all factor into platform selection. A lightweight survey drone may thrive with four rotors, while a high-altitude mapping platform or disaster response unit may require six or eight for added thrust and survivability.

Beyond these standard formats, engineers have pushed the boundaries even further with specialized multirotor designs. Dodecacopters—aircraft with twelve independently powered rotors—have been developed for ultra-heavy lift missions or synchronized swarm behavior where redundancy is paramount. Coaxial rotor systems, featuring stacked rotors spinning in opposite directions on the same axis, allow for compact frames with high thrust, often used in confined or urban environments where traditional rotor arms would be impractical. Hybrid VTOL platforms combine multirotor lift with fixed-wing flight capability, transitioning from vertical takeoff to efficient forward motion, and offering greater range and speed for long-haul or point-to-point missions.

These specialized systems are now entering service in roles once reserved for manned aircraft or ground-based infrastructure. Military logisticians use them to resupply forward units. Urban air mobility researchers experiment with them for future air taxi concepts. Swarm operations—coordinated by AI-driven algorithms—envision flocks of micro- or mid-sized multirotors working in concert for surveillance, search and rescue, or electronic warfare.

The evolution of multirotor configurations reflects a larger truth about drone technology: there is no one-size-fits-all solution. Instead, the field continues to diversify, adapting form to function, and enabling a new era of tailored aerial systems built not only to fly—but to perform with purpose across an ever-expanding range of applications.

Uncrewed aerial vehicles are now central to modern asymmetric warfare and hybrid conflicts, reshaping how smaller nations and irregular forces project power. Nowhere was this more evident than in the 2020 Nagorno-Karabakh war, where Azerbaijan employed the Turkish-manufactured Bayraktar TB2 drone with devastating precision against Armenian defenses. In a conflict that had long been defined by entrenched artillery and static frontlines, the TB2 introduced a new dynamic. Its combination of affordability, endurance, and autonomous targeting helped Azerbaijan achieve rapid battlefield dominance. Fixed targets, mobile armor, and even anti-aircraft systems were neutralized by drone strikes, demonstrating how low-cost UAVs could deliver strategic results once reserved for advanced air forces.

The same technologies revolutionizing conflict zones were also finding their way into the cosmos. NASA, long a pioneer in autonomous systems, embraced drone technology for interplanetary exploration. In 2021, the Ingenuity helicopter became the first aircraft to achieve powered, controlled flight on another planet. Operating in the thin atmosphere of Mars, Ingenuity performed dozens of successful flights over a multi-year mission, proving that aerial reconnaissance could be accomplished beyond Earth. Its compact, solar-powered rotorcraft design was guided by onboard autonomy and inertial navigation—tools derived directly from terrestrial drone innovation.

NASA's ambitions did not stop at Mars. Set to launch in 2027, the Dragonfly mission will deliver a nuclear-powered aerial drone to Saturn's moon Titan, where it will explore vast, uncharted landscapes rich in organic compounds and methane lakes. Arriving in the 2030s, Dragonfly will use quadcopter-style rotors to traverse Titan's low-gravity, dense atmosphere—collecting data on surface chemistry, geology, and the potential for prebiotic life. With these missions, UAVs have transcended their Earthbound military and commercial origins, entering the realm of planetary science and deep-space exploration.

Back on Earth, the integration of artificial intelligence, advanced sensors, and energy-efficient propulsion systems has enabled the development of drone swarms—networks of small UAVs operating collaboratively, often with minimal or no human oversight. These swarms offer formidable tactical and logistical advantages. Capable of real-time mapping, area denial, target saturation, and adaptive response, drone swarms are envisioned as both offensive tools and support assets in future battlespaces. Their ability to overwhelm defenses, navigate cluttered environments, and coordinate autonomously makes them uniquely suited to urban warfare and electronic countermeasure operations.

According to GlobalData, the global market for military uncrewed aerial systems is projected to grow at a compound annual growth rate of 4.8 percent over the next decade. From an estimated $12.5 billion in 2024, the market is expected to reach nearly $20 billion by 2034. This growth reflects a convergence of defense priorities, technological acceleration, and shifting geopolitical demands. Armed forces around the world are investing not only in drone fleets, but in the AI systems, cyber infrastructure, and logistics chains that support them.

Agencies like the Defense Advanced Research Projects Agency (DARPA) continue to drive these innovations. From autonomous swarms to GPS-denied navigation, DARPA's research has pushed the limits of what uncrewed systems can achieve. Programs focused on AI-enabled target recognition and nano-UAVs for indoor reconnaissance exemplify the agency's commitment to systems that operate faster, smarter, and more independently than their predecessors. One conceptual pivot has been the move from "human-in-the-loop" to "human-on-the-loop" control—placing operators in supervisory roles while letting machines execute time-critical decisions autonomously.

DARPA's investments laid the groundwork for advanced platforms such as the X-47B, a stealth UAV capable of autonomous aerial refueling and carrier-based operations. The Gremlins program pushed the envelope further, envisioning drone swarms that could be launched and recovered in mid-air from transport aircraft like the C-130. These initiatives are not just technical demonstrations—they represent a fundamental rethinking of how airpower is projected in contested environments.

While military UAVs dominated headlines, the commercial and civilian drone markets quietly exploded. By the mid-2010s, improvements in lithium-polymer battery chemistry, lightweight carbon-fiber composites, and open-source flight controllers allowed individuals and startups to build and operate sophisticated drones at a fraction of previous costs. The launch of the DJI Phantom in 2013 marked a turning point. With integrated GPS, stabilized cameras, and intuitive flight controls, it brought professional aerial photography within reach of consumers and small businesses. It also sparked a wave of drone startups across agriculture, construction, cinematography, and surveying.

In 2016, the FAA's Part 107 regulations created a legal framework for commercial drone operations in U.S. airspace, enabling licensed pilots to perform aerial work for profit. This regulatory milestone spurred a wave of adoption across industries. Farmers used drones for crop monitoring and precision spraying. Construction firms employed them for volumetric surveys and progress documentation. Emergency responders deployed UAVs to search for survivors in collapsed buildings, assess damage after hurricanes, and monitor wildfires in real time.

Companies like Amazon, UPS, and Zipline began testing drone-based logistics and medical delivery. In Rwanda and Ghana, Zipline's fixed-wing drones delivered blood and vaccines to remote clinics. In the U.S., fire departments used quadcopters to scan hazardous sites, while police agencies experimented with drones for crowd monitoring, suspect tracking, and traffic assessment. What had started as a military innovation had become a global toolset—reshaping not only how wars were fought, but how goods moved, cities were managed, and disasters were addressed.

By 2020, the global drone industry had reached an estimated value of over $20 billion, a figure that continued to climb as commercial applications and dual-use technologies expanded across sectors. What began as a niche within military research had become a global enterprise, influencing everything from national defense to environmental conservation, supply chain logistics to entertainment. The unmanned aerial vehicle had matured—not just as a platform, but as a foundational component of modern technological infrastructure.

The contemporary UAV landscape is no longer defined solely by the military. Instead, it is a sprawling, multi-sector ecosystem driven by the convergence of hardware, software, regulation, and real-time data. Defense contractors such as General Atomics, Northrop Grumman, Boeing, and Elbit continue to lead in the development of large-scale strategic drones for reconnaissance, strike, and electronic warfare. But they now share the stage with commercial innovators like DJI, Skydio, Parrot, and Autel—companies producing compact, agile drones for consumer, industrial, and enterprise markets. These firms are setting standards in autonomy, obstacle avoidance,

AI navigation, and user experience, often outpacing traditional aerospace giants in development speed.

Regulatory bodies have also grown in stature and influence. Agencies like the Federal Aviation Administration (FAA), the International Civil Aviation Organization (ICAO), and national aviation authorities play an increasingly pivotal role in shaping the airspace policies and operational limits of UAVs. Their decisions govern where and how drones may fly, what data can be collected, and how safety, privacy, and accountability are maintained in crowded skies. These frameworks are evolving rapidly, trying to keep pace with a technology that is no longer experimental, but essential.

Complementing the hardware and regulatory infrastructure is a rapidly expanding digital layer—software and data services that transform raw UAV output into actionable intelligence. From real-time video analytics and AI-driven object recognition to cloud-based fleet management and mission planning, this data ecosystem is critical to unlocking the full potential of UAVs. Companies specializing in drone analytics, cybersecurity, and autonomous coordination are now as vital to the industry as airframe manufacturers.

This convergence reflects a larger truth: drones have moved from the fringes of aerospace innovation to its very center. They are no longer exotic. They are foundational. Once defined by reconnaissance missions and battlefield strikes, UAVs now touch nearly every facet of contemporary life. They deliver medical supplies in remote villages, track wildlife migrations across national parks, inspect powerlines and wind turbines, scan disaster zones for survivors, and transport parcels across congested cities. They film movies, monitor crops, survey mines, and enforce borders.

UAVs are ubiquitous, indispensable, and evolving faster than any aviation technology in history. The sky is no longer a barrier—it is infrastructure. And in this airborne domain, drones have become the most agile and adaptive agents of change. Their story is not just one of flight, but of transformation—how a once-radical concept became one of the most powerful tools of modern civilization.

Chapter 2: Hunters and Watchers – Drones on the Post-9/11 Battlefield

The dawn of the 21st century ushered in a seismic shift in the nature of conflict and the tools used to fight it. The September 11, 2001, terrorist attacks marked a strategic turning point in U.S. military doctrine. Confronted by a shadowy enemy with no borders, no uniforms, and no fixed infrastructure, the United States turned to technology to project power with precision—and discretion. At the forefront of this evolution stood a tool that would come to define modern asymmetrical warfare: the unmanned aerial vehicle.

From the tribal regions of Pakistan to the deserts of Yemen and the mountains of Afghanistan, drones offered an unprecedented combination of persistent surveillance, surgical strike capability, and political deniability. They could loiter for hours over suspected insurgent hideouts, track targets across national boundaries, and strike with little warning—without placing American personnel in harm's way. At the heart of this revolution were two aircraft that changed the face of warfare: the MQ-1 Predator and its more powerful successor, the MQ-9 Reaper. Operated remotely by ground crews thousands of miles away, these drones transformed intelligence gathering and targeted killing into scalable, repeatable, and legally ambiguous tools of national power.

Uncrewed aerial vehicles, while often resembling their manned counterparts in silhouette, diverge sharply in design philosophy. The absence of a human pilot fundamentally alters the engineering equation. Without the need for a cockpit, flight instruments, environmental controls, life support, or crash survivability systems, UAVs can be built lighter, leaner, and with mission-specific optimization. Every component—from fuselage to avionics—is tailored for performance, not pilot safety.

This shift in priorities enables a wide range of UAV configurations. Small surveillance drones used by military units or recreational users may carry only a lightweight camera and transmitter, requiring a fraction of the lift and structural mass of a crewed aircraft. Even larger military UAVs, capable of carrying laser-guided munitions, are often lighter than crewed aircraft with similar payloads. Freed from the weight penalties of armor plating, ejection seats, or redundant hydraulics, they prioritize endurance, climb rate, and stealth over crew survivability.

Civilian UAVs, especially quadcopters and multirotors, have capitalized on this design freedom. Their unconventional layouts—multiple vertically aligned rotors for lift and stabilization—would be impractical in manned flight due to range and efficiency constraints, but they are perfectly suited for short-range tasks like filming, surveying, or search-and-rescue missions.

The result is an explosion of aerial platforms that operate in three dimensions without needing the infrastructure of a runway or a trained pilot onboard.

This evolution has also unlocked extraordinary feats of endurance. Over the past several decades, a number of UAVs have set world records for flight duration, demonstrating capabilities once thought unreachable by conventional aircraft. In 2001, the U.S. Air Force's RQ-4 Global Hawk flew non-stop from California to Australia, covering more than 7,000 miles in a single mission. In 2010, the Zephyr—an ultra-lightweight solar-powered drone developed by QinetiQ—remained aloft for over two weeks without landing, showcasing the potential of UAVs for high-altitude, long-endurance (HALE) missions. More recently, experimental systems have combined solar panels, high-efficiency batteries, and autonomous navigation to enable persistent flight lasting weeks or even months—ideal for communications relay, environmental monitoring, or strategic reconnaissance.

These milestones reveal a deeper truth about unmanned flight. When aircraft are liberated from the physical and psychological constraints of human passengers, they can be pushed to extremes in endurance, altitude, stealth, and precision. From battlefield to stratosphere—and now, even to other planets—UAVs are redefining what it means to fly.

By the close of the 20th century and into the early 21st, UAVs had begun to demonstrate endurance capabilities that rivaled or exceeded their manned counterparts. In 1989, Boeing unveiled the Condor, a high-altitude reconnaissance drone that remained aloft for 58 hours and 11 minutes—a remarkable feat at the time. Built for strategic surveillance, the Condor was an early testament to the potential of uncrewed endurance platforms and is now preserved at the Hiller Aviation Museum as a symbol of UAV innovation.

Three years later, in 1992, General Atomics showcased the Gnat, a lighter, more flexible UAV that managed a 40-hour continuous flight. Though less well-known than its successor, the MQ-1 Predator, the Gnat played a foundational role in convincing U.S. defense planners that long-duration UAV surveillance was viable, cost-effective, and scalable.

Not all endurance records came from military giants. On August 11, 2003, a small hobbyist-class UAV called the TAM-5 became the smallest uncrewed aircraft ever to cross the Atlantic Ocean. Weighing just a few pounds, it flew nearly 39 hours to complete the journey—a vivid demonstration of what could be achieved with minimal payload and maximum efficiency.

Around the same time, in the United Kingdom, QinetiQ was developing a different kind of endurance platform: solar-electric UAVs capable of harvesting energy from the sun. In 2007, their Zephyr drone stayed airborne for 54 hours. One year later, it shattered its own record by remaining aloft for 82 hours and 37 minutes. These aircraft proved that solar-powered flight could be more than a novelty—it could be a foundation for persistent aerial presence.

That same year, the U.S. Air Force's RQ-4 Global Hawk set an endurance record for operational military drones, flying for 33 hours and 6 minutes without refueling. Built for high-altitude strategic surveillance, the Global Hawk was a leap forward in both range and autonomy.

Then in July 2010, QinetiQ's Zephyr 7 pushed the frontier further. Between July 9 and July 23, it remained airborne for over 14 days—336 continuous hours—while also filing for a record altitude of 70,740 feet. With that mission, the Zephyr entered near-space operations, occupying an altitude band once accessible only to spy planes and satellites.

The endurance bar continued to rise. In 2023, Airbus's latest Zephyr model reached 70,000 feet and completed a 64-day flight, with a long-term goal of 200 days aloft. With these capabilities, Zephyr is no longer just a drone—it is a high-altitude pseudo-satellite, or HAPS. Positioned in the stratosphere, HAPS platforms offer many of the advantages of satellites—persistent coverage, wide-area monitoring, and communications relay—without the cost or rigidity of orbital deployment.

Another promising entry in this class is the PHASA-35, developed in the UK. Operating at 65,000 feet, it demonstrated station-keeping for 24 hours and remains in advanced testing. Though not yet deployed at scale, PHASA-35 represents a new breed of uncrewed aircraft tailored to upper atmospheric conditions—where air is thin, temperatures plunge, and solar radiation is intense.

These record-setting missions highlight more than just technical endurance. They point to a strategic shift: UAVs are increasingly being used for persistent intelligence, disaster monitoring, climate research, and high-altitude communications—all missions once the exclusive domain of satellites. As battery chemistry, solar efficiency, and lightweight materials continue to improve, the operational ceiling and flight duration of UAVs are expected to expand further, redefining the possibilities of aerial presence on Earth—and beyond.

At the heart of these capabilities lies a fundamental difference in control architecture. Because UAVs lack onboard pilots, their control systems must bridge distance, latency, and autonomy in ways manned aircraft never faced. Instead of cockpit windows, operators rely on live video feeds streamed from onboard cameras. Commands and telemetry are transmitted via radio frequency links, often through satellite relays. Flight control inputs—once manual and tactile—are now digital and scripted, with ground stations issuing high-level instructions while onboard computers handle execution.

Autopilot software plays a central role in UAV operation, managing everything from basic stabilization to full-route planning and autonomous return-to-base functionality. While autopilot systems exist in manned aviation, UAV implementations are typically more deeply integrated with mission-specific payloads, such as cameras, radar systems, or communication relays. The result is an aircraft that flies itself in most respects, checking in with human operators only when needed.

Freed from the limitations of human physiology—no cockpit, no oxygen systems, no ejection seats—UAV designers can adopt airframe configurations tailored to mission, not manpower. Some UAVs, like flying wings, are optimized for stealth and efficiency. Others, like blended wing bodies, maximize internal payload space while minimizing drag. Conventional fuselage-plus-tail designs offer maneuverability and modularity, ideal for long-range reconnaissance or multi-mission roles. In the civilian space, quadcopters and other multirotor platforms dominate, thanks to their simplicity, control precision, and ability to hover in tight spaces.

Each configuration reflects a deliberate trade-off: stealth versus lift, endurance versus maneuverability, complexity versus redundancy. And with no pilot onboard, those trade-offs are no longer bound by human tolerance, but by technological ambition.

Ornithopters – Flapping-Wing Propulsion

Among the most exotic forms of uncrewed aerial systems are ornithopters—drones that fly not by spinning blades or fixed wings, but through the flapping of their wings, mimicking the mechanics of birds or insects. Though the idea of flapping-wing flight dates back to ancient times, it has only become technically viable in recent decades, thanks to breakthroughs in miniaturization, lightweight materials, and precision actuation. Today, ornithopters represent a niche but promising class of UAVs with unique aerodynamic properties and mission applications.

Unlike fixed-wing aircraft, which rely on steady airflow to generate lift, or rotary-wing drones, which spin blades to hover, ornithopters produce both lift and thrust through oscillatory wing motion. This motion can be achieved using compact electric motors, piezoelectric actuators, or even experimental artificial muscles that flex and contract like biological tissue. The result is a drone that moves through the air in a way that closely resembles the flapping flight of living creatures—an advantage that is not merely aesthetic but operational.

Ornithopters are typically divided into two scale categories. Macro-scale designs emulate birds, with broad, flexible wings and relatively slow wingbeats. These are often used for longer-distance reconnaissance and can carry lightweight imaging equipment. Micro- and nano-scale ornithopters, by contrast, draw inspiration from insects, with rapid wing oscillation and high maneuverability. At this scale, designers face unique aerodynamic challenges, including nonlinear airflow, rapid stall recovery, and unsteady vortex generation—all of which are also found in natural insect flight.

One of the most compelling features of ornithopters is their potential for stealth. A UAV that looks and sounds like a bird or a large insect can blend into both natural and urban environments with minimal detection risk. Unlike quadcopters, which produce a distinctive whine, or fixed-wing drones, which have a predictable flight profile, ornithopters exhibit organic, irregular motion and low acoustic signatures. Their natural mimicry provides a form of passive camouflage that can evade both human observation and basic radar detection.

This makes flapping-wing UAVs highly attractive for missions where discretion is paramount. They are being explored for use in espionage, urban surveillance, reconnaissance in denied environments, and covert intelligence gathering. A bird-shaped drone perched on a windowsill may arouse no suspicion. An insect-like micro-UAV drifting into a conference room could gather audio or visual intelligence undetected. These capabilities are being actively researched by military and intelligence agencies, as well as academic institutions focused on biomimetic engineering.

Though still largely experimental and constrained by power supply and flight duration limitations, ornithopters symbolize a broader shift in UAV development—from brute-force mechanical systems to biologically inspired flight. They are not designed to compete with high-altitude surveillance platforms or long-range delivery drones, but to occupy a different kind of airspace altogether—one where stealth, subtlety, and adaptability are the mission's most critical parameters.

Micro-Ornithopters and Insect-Inspired Flight

As the quest for miniaturization accelerates, a new frontier of aerial robotics has emerged: micro-ornithopters—flapping-wing drones that weigh less than a gram and take direct inspiration from the flight mechanics of insects and small birds. These micro aerial vehicles (MAVs) push the boundaries of what is aerodynamically and mechanically possible at small scales, using ultra-light materials, artificial muscle actuators, and novel control algorithms to replicate the agility and efficiency of biological flight.

Several advanced research programs have yielded remarkable breakthroughs. Some prototypes, inspired by flies, have demonstrated the ability to perch on vertical surfaces—an achievement long considered a milestone in aerial robotics due to the extreme precision and energy efficiency required to decelerate, position, and anchor mid-flight. Others emulate beetles, deploying shell-like wing covers (elytra) that not only protect delicate components during landing but also contribute to lift during takeoff.

These insect-mimicking MAVs are often powered via lightweight tethers in laboratory settings, allowing for continuous energy supply and real-time control during development. But untethered models are beginning to emerge, incorporating solar microcells or ultra-light batteries for limited autonomous operation. Wing motion is typically driven by advanced actuators, such as dielectric elastomers or carbon nanotube yarns, which simulate the contraction and flexion of insect muscles. With these technologies, engineers have crafted drones that hover, dart, and pivot in ways indistinguishable from their natural counterparts.

Some prototypes are capable of navigating tight indoor spaces, maneuvering through cluttered forests, or performing controlled landings on walls and ceilings—mirroring the behavioral repertoire of wasps, beetles, or dragonflies. Their ability to traverse confined or chaotic environments makes them ideal for urban reconnaissance, cave exploration, or battlefield surveillance in dense terrain. A swarm of such MAVs could sweep through a collapsed building, map interiors, or relay environmental data without risking human responders.

Yet for all their promise, micro-ornithopters face formidable technical challenges. Their flapping mechanisms, while biologically elegant, are generally less energy-efficient than conventional propellers. Maintaining stable flight requires complex articulation across multiple axes, demanding advanced onboard stabilization software and highly responsive control systems. Payloads remain minimal—most MAVs can carry only the lightest cameras or sensors, and reliable communication links are difficult to maintain over distance without external support.

Despite these constraints, their tactical and scientific advantages are undeniable. Their biological mimicry grants them ultra-low visibility in both visual and acoustic spectra. Their size allows them to enter spaces inaccessible to larger drones. And their natural flight patterns enable them to operate in settings where mechanical drones would stand out. In military contexts, they represent a new class of stealth asset—unobtrusive, expendable, and effective. In scientific fields, they offer tools for pollination studies, environmental monitoring, and indoor aerial mapping on a microscale.

As actuator technology advances and energy storage becomes more efficient at small scales, micro-ornithopters are poised to become more than experimental novelties. They are evolving into real-world tools—tiny, autonomous agents of flight that extend the reach of both surveillance and science into places no other aircraft can go.

The Cold War and Strategic Reconnaissance: Enter the D-21

The D-21 drone was born in the shadow of one of the most advanced aircraft ever flown—the CIA's A-12 OXCART, a Mach 3 reconnaissance platform developed in secrecy as a precursor to the SR-71 Blackbird. The A-12 was already a technological marvel, capable of cruising above 90,000 feet at over three times the speed of sound. At those altitudes and velocities, it was virtually invulnerable to Soviet air defenses of the time. But even that wasn't enough to eliminate risk.

Concerned about the consequences of losing a manned aircraft over hostile territory, the CIA sought an alternative: an uncrewed version of the A-12 that could carry out deep-penetration reconnaissance without risking American lives—or political fallout. The result was a bold new aircraft unlike anything the world had seen: a high-altitude, high-speed drone designed for autonomous surveillance missions across denied airspace.

Lockheed's Skunk Works team, led by legendary aircraft designer Kelly Johnson, responded with the Q-12. Sleek, dart-like, and powered by a Marquardt RJ43-MA-11 ramjet engine, the aircraft was optimized for autonomous flight at extreme speeds and altitudes. Capable of reaching Mach 3.5 and operating above 90,000 feet, the Q-12 could follow a pre-programmed flight path over enemy territory, collect photographic intelligence, and return without ever needing a pilot onboard.

The aircraft was later redesignated the D-21—"D" for "daughter," a nod to its origins from the A-12 "mother" program. But developing such an exotic platform required an equally innovative launch system. Rather than attempting ground launch with a rocket booster or risking air-to-air deployment at suboptimal speeds, Johnson's team devised a daring solution: mounting the D-21 directly atop a modified A-12, designated the M-21.

The M-21 featured a reinforced fuselage and a second cockpit, offset behind the pilot, for a launch control officer. Between its twin vertical stabilizers, the aircraft carried the D-21 in a streamlined cradle. At speeds exceeding Mach 3 and altitudes above 80,000 feet, the drone would be jettisoned. Once released, its ramjet engine would ignite, allowing the D-21 to surge forward on its mission.

The drone's onboard camera system captured high-resolution images using film, not digital sensors—technology that remained classified at the time. After completing its flight path, the D-21 ejected a sealed film capsule, which descended under parachute into international waters. Recovery was attempted midair using a JC-130 Hercules equipped with a trailing hook system—a technique adapted from earlier Corona satellite film capsule recoveries. If the capsule missed midair capture, ships and helicopters stood ready to retrieve it before it sank or was intercepted.

In theory, it was a brilliant concept: autonomous intelligence collection at hypersonic speed, immune to interception and deniable if compromised. In practice, it proved far more dangerous than anticipated. Would you like to continue with the next section on the M-21 accident and the eventual pivot to B-52 launches under Tagboard?

· The D-21 was never designed to return. After completing its reconnaissance mission and ejecting its film capsule, the drone would either self-destruct or crash into remote, unpopulated terrain. Its one-way nature reflected both its mission philosophy and the limitations of recovery at such extreme speeds and altitudes. What mattered was the film—not the aircraft.

· Flight testing of the D-21 system was conducted at one of the most secretive locations in the world: Groom Lake, Nevada, better known today as Area 51. Under tight CIA and U.S. Air Force security, a team of specialists carried out one of the most ambitious drone development programs of the Cold War. Oversight came from the CIA's Special Projects Team, whose members included engineers, flight test analysts, and telemetry experts. Among them was Thornton D. Barnes, a key figure who worked directly with systems tracking, data analysis, and test evaluation for the D-21's advanced propulsion and guidance architecture.

From the beginning, the program encountered serious challenges. The unique mating of the D-21 to the M-21 carrier aircraft created aerodynamic and structural stresses that had never been tested at such speeds. Early test flights experienced multiple separation failures, where the drone either failed to release cleanly or recontacted the mother aircraft in dangerous proximity. Timing the ignition of the ramjet engine was another obstacle. The transition from passive glide to autonomous cruise required precise sequencing, and any delay or mistiming could cause catastrophic instability. Structural integrity was also an ongoing concern. The thin, dart-like airframe of the D-21 was designed for efficiency, not durability, and the extreme heat and vibration of Mach 3+ flight pushed materials and components to their limits.

Each test flight brought new data and new complications, forcing the team at Groom Lake to continuously refine procedures, strengthen materials, and adapt launch protocols. In a program where even minor miscalculations could lead to fiery failure at unimaginable speeds, the D-21 proved to be one of the most complex and dangerous drone systems ever flown. But it was also one of the most advanced—a glimpse into the unmanned future of reconnaissance and strategic aviation.

One of the most tragic and consequential moments in the D-21 program occurred on July 30, 1966. During a high-speed test off the coast of California, the D-21 was launched from its M-21 mothership as planned. But almost immediately after separation, the drone suffered an engine unstart—a violent disruption of airflow through the ramjet—causing it to veer back into the carrier aircraft. The collision destroyed both vehicles in midair. The pilot, Bill Park, managed to eject and survived the incident. Tragically, launch control officer Ray Torrick also ejected but drowned after landing in the Pacific Ocean before rescue teams could reach him.

The loss of Torrick and the destruction of both the drone and the highly specialized M-21 marked a turning point. The CIA and Lockheed deemed the dual-aircraft launch system too risky to continue. The M-21 program was immediately terminated, and the focus shifted to a safer, more conventional launch platform: the B-52H Stratofortress.

To accommodate the change, Lockheed engineers reconfigured the D-21 as the D-21B. The updated model eliminated the need for high-speed aerial release. Instead, the drone was mounted under the wing of a modified B-52 and launched using a powerful solid rocket booster that carried it to ramjet ignition speed and altitude. This new phase of the program was assigned the codename Tagboard and would become one of the most secretive drone operations of the Cold War.

Between 1969 and 1971, four operational D-21B missions were flown over the People's Republic of China. The primary target was Lop Nor, a remote region in Xinjiang where China conducted its nuclear weapons tests. The goal was to obtain high-resolution photographic intelligence on nuclear yields, delivery systems, and facility expansion—data the United States could not reliably collect using satellites alone.

Each mission, however, met with failure. On the first flight, the D-21B crashed shortly after release from the B-52. The second mission reached its target and successfully dropped the film capsule, but the payload landed deep in inaccessible territory and could not be recovered. The third flight suffered a systems malfunction and failed to release the film capsule entirely. The fourth and final mission returned its capsule successfully, only for analysts to discover that the film contained no usable imagery.

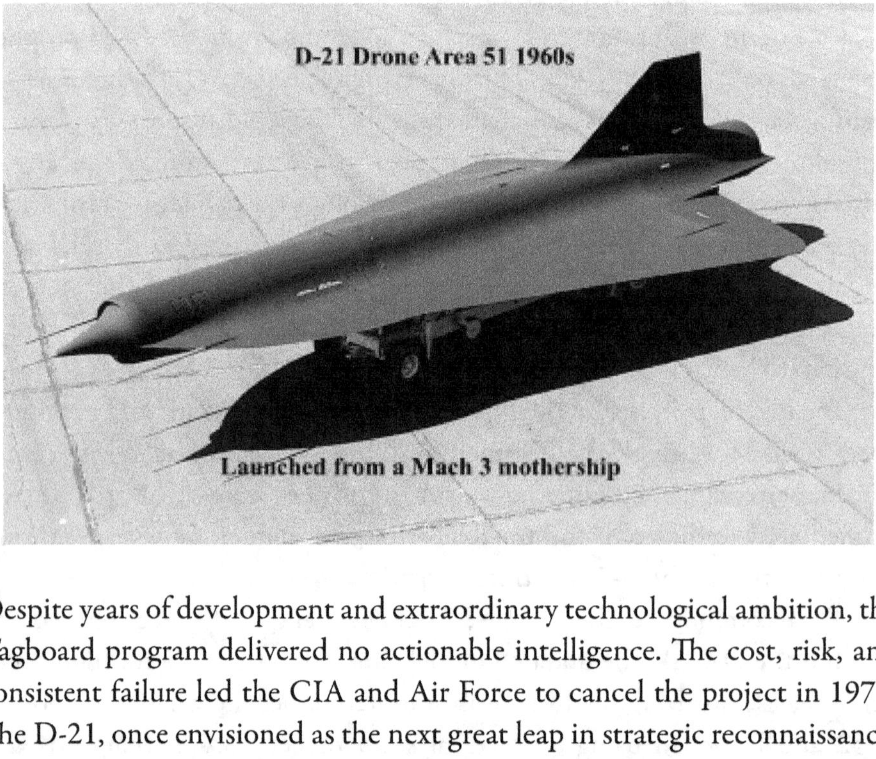

Despite years of development and extraordinary technological ambition, the Tagboard program delivered no actionable intelligence. The cost, risk, and consistent failure led the CIA and Air Force to cancel the project in 1971. The D-21, once envisioned as the next great leap in strategic reconnaissance, was quietly retired. Several surviving airframes were placed in storage or sent to museums, and the program itself remained classified for years.

Yet in retrospect, the D-21 was not a failure. It was a harbinger. It demonstrated that autonomous, high-speed surveillance platforms could operate at the edge of Earth's atmosphere—without a pilot and with minimal visibility. Its challenges helped inform future developments in drone propulsion, guidance, and recovery. And its legacy would echo decades later in the rise of strategic unmanned systems capable of reaching places no human crew could go.

Despite the frustrations and failed recoveries, the D-21 program left behind a powerful truth: an unmanned platform could penetrate denied airspace at altitudes and velocities no crewed aircraft could match. Flying above 90,000 feet at over Mach 3, the drone had proven it could go where satellites struggled to see and pilots could not survive. Its flights, though operationally flawed, foreshadowed a future where machines—not men—would assume the riskiest roles in strategic reconnaissance.

By 1971, however, both the CIA and the U.S. Air Force reached the same conclusion—the D-21 program was no longer cost-effective. Advancements in satellite surveillance had dramatically shifted the intelligence landscape. New space-based systems like the KH-8 *Gambit* and KH-9 *Hexagon* offered broader geographic coverage, superior reliability, and none of the geopolitical risks of a drone crashing in hostile territory. In a changing technological and strategic environment, the D-21 had simply outlived its moment.

The program was quietly terminated. Remaining airframes were either stored in classified warehouses or later transferred to museums, their history declassified in fragments over the decades. But the legacy of the D-21 far exceeded its limited operational run.

It had proven that autonomous, high-speed drones could operate far beyond line-of-sight or real-time control—a principle that would become essential for both strategic and tactical UAVs in the decades to follow. Its film-recovery capsule system, though ultimately unreliable, helped validate techniques that informed satellite retrieval efforts. And it forced engineers to confront a host of difficult challenges: separation dynamics at Mach 3, heat shielding for sustained hypersonic cruise, inertial guidance in GPS-denied environments, and compact propulsion systems optimized for atmospheric edge operations. These were not just academic hurdles—they were the very problems that would define the next generation of unmanned and hypersonic platforms.

The D-21 was a machine ahead of its time. Constructed from titanium, powered by a ramjet, and directed by an inertial navigation system, it embodied the leading edge of 1960s aerospace engineering. But it was also a creature of Cold War urgency—conceived in secrecy, tested in silence, and ultimately shelved not because of what it lacked, but because the world around it had already begun to change. Its form may have faded, but its influence persisted.

Elements of its design and philosophy would later reemerge in aircraft like the RQ-170 *Sentinel*, which mirrored the D-21's emphasis on stealth, altitude, and autonomy. But perhaps the D-21's most important legacy was cultural: it shifted drones from being expendable curiosities to indispensable tools of national strategy. These were not just experimental gadgets—they were platforms entrusted with missions too dangerous, too provocative, or too politically explosive for human pilots to undertake.

And behind every machine stood the minds that built them. While aerospace history is often shaped by institutions—government agencies, contractors, and classified programs—it is driven forward by individuals. Visionaries whose designs, leadership, or sheer persistence pushed the boundaries of what flight could mean. The rise of unmanned aerial systems was not an accident. It was the result of deliberate, often obsessive work by engineers, military leaders, and theorists who saw beyond convention and embraced a new paradigm in aviation.

Few individuals loomed larger in that history than Clarence "Kelly" Johnson, the brilliant and famously demanding chief designer of Lockheed's Skunk Works. Johnson led the teams that created the U-2 spy plane, the SR-71 *Blackbird*, and the A-12 *OXCART*—aircraft that redefined reconnaissance and performance standards for an entire generation. He was also instrumental in the D-21's development, overseeing a program that extended the logic of stealth and speed into the unmanned realm.

Johnson was among the first aerospace leaders to fully embrace uncrewed flight as a necessity in intelligence operations. He understood that the Cold War was not just a battle of arsenals—it was a battle for deniability, persistence, and technological superiority. He saw drones not as disposable novelties, but as mission-critical instruments of national power. His signature design philosophy—*quick, quiet, and quality*—became the cultural DNA of the Skunk Works and set the benchmark for every classified aerospace program that followed. It was under his watch that the impossible became routine and that the age of the pilotless aircraft was born not in theory, but in titanium.

CIA Project Aquiline – Pioneering UAV Espionage at Area 51

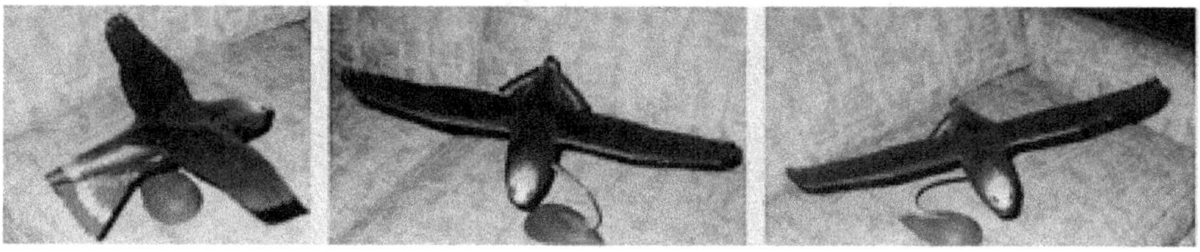

In the shadowy world of Cold War reconnaissance, where U-2 overflights and satellite programs like *Corona* captured the public imagination, a lesser-known but visionary effort unfolded deep within the CIA's Directorate of Science & Technology. Launched in the late 1960s, *Project Aquiline* represented the Agency's bold attempt to create a stealthy, long-range unmanned aerial vehicle capable of infiltrating the most heavily defended airspace on Earth—without risking a pilot or sparking an international crisis.

The project emerged in the aftermath of the 1960 U-2 incident, in which CIA pilot Francis Gary Powers was shot down over Soviet territory, setting off a geopolitical firestorm. Determined to avoid future diplomatic disasters, the CIA sought an alternative: a drone that could travel thousands of miles, collect high-value intelligence, and return home—unnoticed, and ideally, intact.

Aquiline was conceived to bridge a critical gap in strategic reconnaissance. It would be small, radar-evading, and almost silent, capable of deep-penetration missions over the Soviet Union, China, Cuba, and other denied territories. Its payloads would include high-resolution film cameras, electronic and signals intelligence collectors, and possibly even miniaturized nuclear power sources for extreme endurance. It would fly autonomously or via encrypted command links, using inertial navigation and emerging satellite guidance systems to reach its target, loiter if needed, and return undetected.

To achieve this, the CIA adopted a radically unconventional design. Aquiline's airframe was shaped to resemble a bird, with a slender fuselage and a wingspan of approximately ten feet. Constructed from composite materials to reduce weight and radar signature, the drone was intended to blend into the sky—both visually and electronically. Agency engineers even proposed giving the drone birdlike wing articulation to mimic natural flight. Some documents suggest the drone was to be disguised as a vulture or other soaring raptor, making it appear innocuous to both ground observers and radar operators.

The propulsion system was equally innovative. Early concepts used a miniature Wankel rotary engine or a small piston engine optimized for long-duration cruise at high subsonic speeds. Modular payload bays allowed for mission-specific equipment, from reconnaissance optics to ELINT antennas. Internal design documents also indicate that a compact radioisotope thermoelectric generator—a nuclear power source—was explored as a long-endurance option, though never deployed in flight tests.

All testing and development were carried out under tight security at Groom Lake, Nevada—Area 51. There, Aquiline joined other black programs, including the A-12 *Oxcart*, the D-21 *Tagboard*, and advanced satellite recovery systems. The CIA's Special Projects Team, working alongside aerospace contractors and classified research centers, handled telemetry, guidance integration, and airframe evaluation. Thornton D. Barnes and other team members played direct roles in tracking Aquiline's test flights and interpreting performance data under conditions that often pushed the limits of existing technology.

Despite promising early results, the program faced formidable obstacles. The miniature engine systems, though efficient, lacked long-term reliability. Navigation systems—precursors to GPS—struggled to maintain precision over vast distances without external updates. Communications links were vulnerable to interference, and the limited payload capacity imposed hard trade-offs between endurance and sensor capability. Inter-agency rivalries between the CIA and the Department of Defense further complicated funding, ownership, and deployment authority.

By the early 1970s, rising costs and shifting technological priorities led the CIA to quietly terminate Aquiline before it reached operational deployment. The intelligence community increasingly leaned on improved satellite platforms like the KH-8 *Gambit* and KH-9 *Hexagon* to meet its strategic imagery needs. Yet the lessons of Aquiline endured. The drone had pioneered concepts decades ahead of their time—stealthy design, autonomous navigation, multi-sensor payloads, and deep-penetration reconnaissance without human risk.

For years, Aquiline remained a footnote, buried beneath layers of classification. It lacked the mythos of the SR-71 or the notoriety of the U-2. But declassified documents and accounts from Area 51 veterans have since shed light on its importance. Aquiline didn't fail—it foreshadowed. Its core ideas reemerged in the strategic UAVs of the 21st century, from the RQ-170 *Sentinel* to the high-altitude Global Hawk and beyond.

Though it never flew in combat, Aquiline marked a turning point in the philosophy of aerial espionage. In the age of satellites and supersonic jets, the CIA had quietly built a drone that flew like a bird, gathered secrets like a spy, and vanished like a ghost.

The Predator: A Hunter Born in the Shadows

The origins of the Predator drone can be traced to a quiet garage in California during the 1980s, where Israeli-born aeronautical engineer Abraham Karem pursued a vision that would one day reshape modern warfare. Karem, a former chief designer for the Israeli Air Force, had immigrated to the United States in the late 1970s. Frustrated by what he viewed as inefficient American drone development, he set out to build a better system—one that could fly longer, watch silently, and return home without ever being seen.

His early efforts materialized in a program called *Amber*, a Cold War-era UAV designed for endurance rather than speed. With lightweight construction and efficient aerodynamics, Amber could loiter over targets for extended durations, offering real-time surveillance capabilities that outperformed anything in the U.S. inventory at the time. The work was conducted under his small firm, Leading Systems Inc., with limited funding and minimal fanfare. But the promise of the platform attracted the attention of military planners and aerospace investors.

Eventually, General Atomics acquired Karem's company and began refining his designs under a new banner. Amber evolved into a more capable and robust platform, eventually designated the RQ-1 *Predator*. With a 55-foot wingspan, a cruising speed of 84 miles per hour, and the ability to stay aloft for more than 24 hours, the Predator marked a major shift in drone philosophy. It wasn't built for stealth or speed—it was built to watch, wait, and endure.

The Predator first entered service in the 1990s, providing real-time imagery during U.S. and NATO operations over the Balkans. In Bosnia, it demonstrated its utility as a persistent eye in the sky, capable of tracking troop movements, monitoring ceasefires, and relaying crucial intelligence. But it was in the years following the September 11, 2001, terrorist attacks that the Predator would evolve from a reconnaissance platform into a weapon of war.

After 9/11, U.S. counterterrorism priorities shifted dramatically, and with them, the mission of the Predator. Armed with AGM-114 *Hellfire* missiles, the drone transitioned from observer to hunter. For the first time in history, an uncrewed aircraft could surveil, identify, and strike a target—all without placing an American pilot in harm's way. This fusion of surveillance and precision lethality ushered in a new era of remote warfare.

In November 2002, the Predator made history when a CIA-operated drone launched a missile that killed al-Qaeda operative Qaed Salim Sinan al-Harethi in Yemen. The strike marked the first publicly acknowledged targeted killing carried out by an armed UAV. Al-Harethi had been linked to the 2000 bombing of the USS *Cole*, and his elimination—executed from the sky by operators stationed thousands of miles away—confirmed what many had only imagined: drones were no longer passive platforms. They were now instruments of national power, capable of delivering justice—or destruction—without warning and without presence.

CIA and JSOC: Parallel Drone Wars

As UAV capabilities matured, the U.S. intelligence and military communities developed two parallel architectures for drone warfare—each shaped by its own legal authorities, organizational culture, and strategic objectives. On one side stood the Central Intelligence Agency, operating under Title 50 of the U.S. Code, tasked with covert action. On the other was the Department of Defense, specifically the Joint Special Operations Command (JSOC), governed by Title 10 and focused on wartime operations.

While both entities flew similar aircraft—Predators, Reapers, and later stealthier platforms—their missions diverged in critical ways. The CIA emphasized intelligence-driven targeting, often spending weeks or months building patterns-of-life dossiers on high-value targets. Strikes were rare, surgical, and frequently unacknowledged. In contrast, JSOC prioritized battlefield tempo. Its drone missions were often rapid, tactical, and executed in support of special operations teams conducting raids in places like Afghanistan, Iraq, Syria, and Somalia.

This bifurcated system created overlapping but distinct drone fleets, targeting lists, and rules of engagement. While collaboration existed, friction was inevitable. Agencies competed for intelligence dominance, operational credit, and strategic influence. To impose structure on an increasingly complex kill chain, the Obama administration formalized a centralized review system known internally as the Disposition Matrix—colloquially, the "kill list."

Each week, intelligence and defense officials presented targeting packages to a national security review board chaired by the President. Photographs, intercepts, behavioral models, and legal justifications were discussed in secure White House meetings. The President personally approved or denied strikes—often selecting individuals for lethal action based on classified intelligence alone. Supporters viewed this as a rigorous safeguard; critics saw it as an unchecked exercise of executive power, bypassing judicial oversight in favor of technocratic warfare.

By the 2010s, drone strikes became a fixture of America's counterterrorism strategy. Operations expanded from Afghanistan and Pakistan to Yemen, Libya, Somalia, and beyond. The targets ranged from senior al-Qaeda leaders to unnamed individuals judged to meet threat criteria through metadata and aerial observation. A new term emerged in national security circles—"signature strikes"—referring to drone attacks based not on identity but behavior. If a person's activities aligned with known terrorist profiles, they could be marked for death without ever being named.

This evolution sparked intense constitutional debate. Civil liberties advocates questioned the legality of targeted killings—particularly those involving U.S. citizens, such as Anwar al-Awlaki. Military strategists debated whether the short-term tactical gains of drone strikes outweighed the long-term strategic risks, including blowback, radicalization, and erosion of international norms.

Yet the machinery of drone warfare kept advancing. By 2020, over 100 countries had integrated UAVs into their armed forces, and at least 17 had operational armed drones. What had started as an experiment in high-altitude surveillance had become the defining instrument of 21st-century warfare—precise, persistent, and politically adaptable.

One of the most controversial dimensions of drone warfare was the rise of what came to be known as "signature strikes." Unlike "personality strikes," which targeted specific individuals whose identities were confirmed through intelligence, signature strikes were predicated on patterns of behavior. If a group of military-aged males were observed congregating near known militant compounds, traveling armed through high-threat regions, or engaging in activities judged suspicious by algorithmic models and analyst review, they could be marked for elimination—without ever being positively identified.

This tactic became a centerpiece of operations in Pakistan's Federally Administered Tribal Areas and in the remote provinces of Yemen, where human intelligence was limited and access for ground forces was nearly impossible. From above, operators watched and assessed. When the observed behavior aligned closely enough with profiles built from previous strikes, the decision to engage could be made within minutes.

But the price was steep. Civilian casualties mounted. Innocent gatherings were sometimes misread. Weddings were mistaken for militant meetings; farmers with shovels were identified as insurgents with rifles. And all of it took place under a veil of secrecy that precluded public explanation or accountability. The resulting outrage—both foreign and domestic—undermined U.S. credibility and sparked widespread debate about the ethics of predictive killing.

The issue reached its most incendiary point in 2011, when a drone strike in Yemen killed Anwar al-Awlaki, an American-born cleric and influential al-Qaeda propagandist. Though many viewed his death as a national security success, it raised a constitutional alarm: al-Awlaki had been executed without trial, judge, or jury. Weeks later, a separate strike killed his 16-year-old son, Abdulrahman—also a U.S. citizen—while the teenager was reportedly searching for his father. The government offered no legal justification, only silence.

The incident sparked a furious national debate about executive authority, due process, and the power to kill American citizens without judicial oversight. Lawmakers demanded answers. Civil liberties groups filed lawsuits. The Obama administration, which had championed transparency and constitutional fidelity, now found itself at the center of a storm that questioned the very legitimacy of remote warfare.

And yet, the war continued—fought not just in distant deserts and mountains, but from secure trailers in Nevada. At places like Creech Air Force Base, drone pilots and sensor operators engaged in lethal combat operations via satellite link, sometimes for 12-hour shifts at a time. Then, they drove home, made dinner, and tucked their children into bed. This bifurcated existence—simultaneously at war and at peace—created a new kind of psychological burden.

Operators spoke of the dissonance: of watching targets for days, observing daily routines, family interactions, and village life through infrared sensors and electro-optical feeds. They witnessed their targets in moments of intimacy, watched children play near their fathers—and were then ordered to strike. The weight of such decisions, made in silence and executed with precision, often returned home with them.

Though drone crews faced none of the physical dangers of traditional combat, they reported emotional trauma on par with deployed troops. Post-traumatic stress, moral injury, and psychological fatigue became common. The human mind, it turned out, was not wired to make life-and-death decisions from behind a joystick—especially not in a world where the battlefield was only one monitor away from suburbia.

In response, the Air Force began to adapt. New mental health programs were established, tailored specifically to the needs of drone personnel. Mission rotations were adjusted to reduce burnout, and support teams were embedded to monitor the psychological toll of remote operations. The drone war, it became clear, was not only altering how battles were fought—it was changing what it meant to be a soldier.

Between 2004 and 2020, U.S. drone operations expanded into a truly global enterprise, stretching across continents and reaching into some of the world's most volatile regions. From the mountains of Afghanistan to the deserts of North Africa, unmanned aerial vehicles became the tip of the spear—delivering intelligence, eliminating high-value targets, and reshaping the battlefield without ever placing a pilot in the sky.

In Afghanistan, drones were omnipresent. From early surveillance flights to armed overwatch missions, they tracked insurgent movements, supported special operations raids, and eliminated senior figures within both the Taliban and al-Qaeda. Predators and Reapers loitered over valleys and compounds for hours, feeding live imagery to commanders and striking with pinpoint accuracy when called upon.

Across the border in Pakistan's tribal regions—particularly North Waziristan—drone strikes became the primary method of targeting al-Qaeda's leadership. Hundreds of covert missions, many flown by CIA-operated UAVs, decimated the network's safe havens. But the intensity of these operations strained the already fragile U.S.-Pakistan relationship. Public protests erupted. Pakistani officials denounced violations of sovereignty. Secret cooperation turned into public condemnation as collateral damage and civilian deaths fueled political backlash.

In Yemen, the drone war took on a hybrid form. Both JSOC and CIA assets were used to pursue al-Qaeda in the Arabian Peninsula (AQAP), a group considered especially dangerous due to its bomb-making expertise and attempts to launch attacks on U.S. soil. The rugged terrain and lack of central governance made traditional operations nearly impossible. Drones filled the void, patrolling the skies above Shabwah and Marib, striking vehicles and compounds suspected of housing militants.

In Somalia, UAVs supported African Union forces and Somali government troops in their long-running struggle against the militant group al-Shabaab. U.S. drones eliminated foreign fighters, bomb-makers, and leadership figures, all while providing real-time surveillance that was instrumental in staging ground raids and counter-insurgency operations.

In Iraq and Syria, drones played a decisive role in Operation Inherent Resolve, the international campaign against the Islamic State. UAVs mapped enemy positions, identified key leadership, and tracked logistical routes. They delivered strikes that dismantled ISIS's command structure and disabled its propaganda and financial networks. From Mosul to Raqqa, drones formed a web of eyes that made hiding from the coalition nearly impossible.

Even in Libya, a country fractured by civil war, drones enabled targeted operations against ISIS holdouts during the Second Libyan Civil War. ISR missions and precision strikes helped dismantle jihadist enclaves near Sirte and Misrata, contributing to temporary stabilization in key regions.

Collectively, these campaigns demonstrated the tactical value and strategic flexibility of UAVs. Drones could be deployed rapidly, operate without basing rights in-country, and deliver both surveillance and firepower without risking lives on the ground. For policymakers, they offered a low-footprint, high-impact solution to a wide range of asymmetric threats.

But the advantages came with consequences. Civilian casualties—sometimes caused by faulty intelligence or misidentification—undermined the legitimacy of drone campaigns. The constant hum of UAVs in the skies above rural villages fostered fear and resentment. Governments accused the U.S. of violating national sovereignty. And in many regions, drone strikes became recruitment tools for the very groups they were meant to suppress.

Human rights organizations and international observers raised persistent concerns: Were legal safeguards adequate? Who was accountable for mistakes? How were targets vetted in countries where intelligence was scarce and local dynamics opaque? The answers were often hidden behind layers of classification, fueling suspicion and debate.

Drones had changed the shape of war—but not always the outcome. While they removed enemies with precision, they sometimes created new ones in the process.

The Reaper and Next-Gen Systems

Introduced in 2007, the MQ-9 *Reaper* represented a quantum leap in unmanned aerial capabilities. Building on the foundation laid by the Predator, the Reaper featured a longer wingspan, greater endurance, and an expanded payload capacity. It could carry up to 3,800 pounds of munitions—including GBU-12 laser-guided bombs and multiple AGM-114 *Hellfire* missiles—transforming it from a surveillance asset into a full-fledged hunter-killer platform. No longer just an eye in the sky, it had become a predator in the most literal sense.

The Reaper was used extensively in Afghanistan, Iraq, Syria, and across the African theater. Operating both under the Department of Defense and in partnership with CIA paramilitary teams, the Reaper performed an array of missions—armed overwatch, target tracking, route clearance, convoy protection, and time-sensitive strikes. By 2020, Reapers had surpassed 2 million flight hours, evolving into not only a weapons platform but a command-and-control node. With upgraded sensors and communications relays, they enabled battlefield coordination and signals intelligence collection in real time.

While the Reaper dominated active warzones, a quieter revolution was underway in stealth and autonomous UAV development. The RQ-170 *Sentinel*, a bat-winged drone operated by the U.S. Air Force and CIA, was designed for high-altitude, low-observable reconnaissance. Its existence was not publicly acknowledged until December 2011, when one crashed nearly intact inside Iran. The incident stunned the intelligence community, revealing that America's most secret drone had not only been deployed in contested airspace but had fallen into enemy hands—provoking a propaganda windfall for Tehran and a setback for U.S. covert operations.

Meanwhile, Northrop Grumman's X-47B pushed the envelope even further. Unlike earlier drones that required human pilots at remote stations, the X-47B was designed to operate autonomously—navigating, taking off, landing, and refueling without direct control. In 2013, it made aviation history by becoming the first UAV to autonomously launch from and land on an aircraft carrier. Later, it successfully completed in-flight refueling operations. Though the program was eventually retired in favor of more cost-effective initiatives, it proved that pilotless aircraft could function across the full spectrum of carrier operations, opening the door to next-generation naval airpower.

As the United States refined its systems, rival powers entered the fray. Russia developed the *Orion* and *Okhotnik-B*; China fielded the *Wing Loong* and *CH-series* UAVs; Iran deployed platforms like the *Shahed-129* and reverse-engineered versions of the captured RQ-170. These countries did not merely observe—they innovated, adapted, and began exporting their own drones to regional allies and proxy forces.

Turkey emerged as one of the most influential drone players of the 21st century. Through its defense company Baykar, it introduced the *Bayraktar TB2*, a medium-altitude, long-endurance UAV that soon became a fixture in conflicts from Libya to Ukraine. The TB2's affordability, precision, and battlefield success earned it widespread acclaim—and concern—among global military analysts. Turkey had not only built an effective drone but a replicable export model, helping reshape the global drone economy.

As these systems proliferated, drones ceased to be niche assets or tools of intelligence agencies alone. They became front-line instruments of conventional war. In Libya, UAVs from Turkey and the United Arab Emirates clashed over Tripoli's skies. In the 2020 Nagorno-Karabakh war, Azerbaijan used Turkish and Israeli drones to destroy Armenian armor, air defenses, and logistics nodes with near impunity. In Ukraine, both sides deployed drones at scale—ranging from commercial quadcopters to loitering munitions—making UAVs an integral component of battlefield doctrine.

The democratization of drone technology had begun. Once confined to superpowers and elite aerospace firms, unmanned systems became modular, affordable, and accessible. Nations that could not afford fighter jets now purchased UAVs capable of both surveillance and strike. Militias and non-state actors began modifying hobbyist drones for combat. The sky had opened not just to innovation, but to disruption.

Israel, one of the earliest pioneers of UAV development, remained a dominant force. Its companies—Israel Aerospace Industries, Elbit Systems, and Aeronautics Ltd.—exported platforms to more than 50 countries, making Israel the second-largest drone exporter after the United States. Systems like the *Heron*, *Searcher*, and the loitering *Harop* offered scalable ISR and precision strike capabilities tailored for asymmetric environments.

But Israel's true innovation was conceptual. Its military had integrated drones into every layer of operations: from strategic deterrence to border patrols, counterterrorism raids, and targeted killings. UAVs were no longer support tools; they were battlefield assets, embedded in doctrine, logistics, and intelligence fusion. For smaller nations looking to amplify their reach without traditional airpower, Israel's model became a blueprint.

In the late 2010s, Turkey emerged as an unlikely leader in the development and deployment of medium-class armed drones. Spearheaded by Baykar Defense and its flagship platform, the Bayraktar TB2, Turkey demonstrated that affordable, modular UAVs could deliver strategic effects across a range of regional conflicts. Where legacy systems had often been prohibitively expensive or entangled in export restrictions, the TB2 offered a different model—low-cost, export-friendly, and combat-proven.

The TB2 quickly gained operational credibility. In Libya, it was used to support the UN-backed Government of National Accord against Khalifa Haftar's Libyan National Army, providing real-time targeting data and executing strikes that turned the tide of battle. In Syria, Turkish forces deployed TB2s in cross-border campaigns against Kurdish militias and Syrian regime positions, using drones to coordinate artillery and airstrikes with precision and speed. But perhaps the most dramatic demonstration came in the 2020 Nagorno-Karabakh war, where Azerbaijan employed Turkish TB2s alongside Israeli loitering munitions to devastating effect—neutralizing Armenian tanks, air defense systems, and troop positions in rapid succession. The conflict served as a textbook example of how UAVs could dismantle conventional ground forces with minimal risk to the operator.

Turkey's drone doctrine sent a resounding message to the world: unmanned systems had leveled the battlefield. With the right mix of sensors, autonomy, and targeting protocols, a mid-sized UAV could decimate an armored brigade. No longer did strategic airpower require fleets of manned jets and complex infrastructure. Drones, launched from a simple airstrip and controlled by a laptop, were now delivering results once reserved for advanced militaries.

Meanwhile, Russia—initially slow to embrace drones beyond tactical reconnaissance—began accelerating its programs during engagements in Syria and Ukraine. Platforms such as the Orlan-10 and Forpost became the workhorses of Russian ISR operations, while the larger Orion UAV was developed for limited strike capabilities. Though Russian drones lacked the sophistication of their Western counterparts, they were integrated effectively into combined arms operations, relaying targeting data for artillery and conducting battlefield surveillance in contested environments.

In Ukraine, Russia's drone efforts took on multiple dimensions. Drones were used not only for reconnaissance but for psychological warfare.

Propaganda units broadcast footage of successful strikes and surveillance clips to sow fear and confusion among enemy ranks. Electronic warfare units leveraged UAVs to jam Ukrainian communications and GPS signals. In some cases, Russian engineers attempted to reverse-engineer downed Western or Israeli UAVs—gleaning what they could from captured wreckage to inform domestic designs.

Though technologically inferior in many respects, Russian drones became formidable when combined with artillery coordination, cyber intrusion, and layered deception campaigns. They embodied a hybrid warfare strategy that fused traditional firepower with information dominance and plausible deniability.

Iran, long accustomed to sanctions and asymmetric warfare, developed a robust and expanding UAV industry to serve both domestic defense needs and its network of regional proxy forces. Platforms like the Shahed-129 and Mohajer series were fielded not only by the Islamic Revolutionary Guard Corps but also distributed to non-state actors across the Middle East. Hezbollah in Lebanon, the Houthis in Yemen, and Shiite militias in Iraq and Syria all received Iranian-made UAVs capable of both surveillance and explosive delivery.

Iranian doctrine embraced the concept of loitering munitions—suicide drones designed to crash into targets and detonate on impact. In 2019, Houthi forces launched a coordinated drone attack against Saudi Arabia's Abqaiq–Khurais oil processing facilities, temporarily knocking out half the kingdom's oil output. The strike, carried out using Iranian-designed drones, stunned international observers with its precision and impact—and underscored the deniability advantage offered by proxy drone warfare.

These developments revealed the shifting center of gravity in drone warfare. No longer monopolized by the U.S. or Israel, UAVs had become tools of influence for mid-tier powers. From Ankara to Tehran, unmanned systems were now instruments of diplomacy, deterrence, and destabilization. They were rewriting the rules of modern warfare—silent, scalable, and lethal.

By 2022 and into 2023, Iran had deepened its role in drone warfare by exporting UAVs directly to Russia. Chief among them was the *Shahed-136*, a loitering munition designed for one-way missions—what came to be known as a "kamikaze drone." Launched in waves, often from truck-mounted platforms, these drones were used extensively in Russian attacks against Ukrainian infrastructure. Their objective was not only tactical disruption, but strategic exhaustion—eroding Ukraine's air defense systems, damaging power grids, and instilling fear among civilian populations. The drones were cheap, plentiful, and difficult to intercept. They revealed a grim truth: attritable unmanned systems could be as destabilizing as precision munitions, especially when used to wage a war of nerves.

China, meanwhile, approached the UAV arms race with industrial force and a deliberate export strategy. Leveraging its growing aerospace sector and state-backed firms, China developed a family of combat-capable drones including the *CH-4 Rainbow*, the *Wing Loong*, and the jet-powered *WJ-700*. These systems offered performance comparable to the American MQ-1 *Predator* and MQ-9 *Reaper*, but at a fraction of the cost—and with fewer export restrictions.

Chinese drones were soon found in arsenals across Asia, Africa, and the Middle East. Pakistan, Nigeria, Egypt, Saudi Arabia, Iraq, and the United Arab Emirates became early adopters. In many cases, these nations had previously been blocked or discouraged from acquiring U.S. drones due to concerns over human rights abuses or political instability. Beijing filled the vacuum, offering not only UAV hardware but also training, maintenance, and loose licensing agreements that allowed for local assembly and customization.

China's strategy was methodical. By dominating the mid-tier UAV export market, it gained access to real-world operational data, expanded its influence among key partners, and chipped away at American drone hegemony. In doing so, it helped normalize drone warfare among non-Western states, shifting the balance of military power in regions where air supremacy had once been out of reach.

In less than two decades, the drone had evolved from a niche tool of counterterrorism into a central instrument of modern statecraft, insurgency, and proxy war. No longer limited to reconnaissance or surgical strikes, UAVs now formed a multilayered architecture of conflict—a flexible system of surveillance, precision, and plausible deniability that could be activated with a satellite link and executed without a single boot on the ground.

The post-9/11 era had not only transformed geopolitics but had also birthed a new warrior class. Operators at remote terminals—once seen as technicians—became combatants in digital battlespaces. These drone crews operated in shadows, launching missiles and guiding loitering munitions from air-conditioned bunkers thousands of miles away. They were data-driven, politically shielded, and able to inflict lethal force without physical presence.

Drones blurred the traditional lines: between war and policing, between surveillance and assassination, between soldier and sensor. In their wake, the rules of engagement bent, and the laws of war strained to keep pace.

As global powers continue to refine drone doctrine—investing in autonomous navigation, real-time data fusion, and AI-driven targeting—the future of warfare is poised to shift again. Not just in altitude or range, but in architecture. The next battlefield may unfold not only in the skies but inside the processors, circuits, and software that animate uncrewed machines. A war of intelligence, of automation, of code—where decisions are made at the speed of algorithms, and the line between human command and machine autonomy grows ever thinner.

Chapter 3: Eyes Above the Homeland – Drones in Policing and Border Security

As drones proliferated on the battlefield, they also began to reshape the domestic landscape—quietly, steadily, and often without public debate. Their integration into civilian law enforcement brought new capabilities but also ignited profound concerns about privacy, surveillance creep, and the potential for constitutional overreach.

The appeal of drones to police departments was straightforward. Compared to helicopters or fixed-wing aircraft, UAVs offered dramatically lower operating costs, faster deployment, and the ability to access places otherwise out of reach—rooftops, alleyways, collapsed buildings, or hazardous crime scenes. They required no pilot's license, consumed no aviation fuel, and could be operated from the back of a patrol vehicle or a laptop in a command van.

By the early 2020s, drones had become commonplace in municipal and regional law enforcement across the United States. Their use expanded into routine and specialized operations alike. In search and rescue missions, UAVs equipped with thermal imaging cameras combed forests, deserts, and urban ruins, locating missing persons in conditions too dangerous or vast for foot patrols. For traffic accident reconstruction, drones captured overhead imagery of crash sites, creating forensic 3D maps that helped investigators model impact scenarios while reducing the time roads remained closed.

At large public gatherings—protests, parades, or political rallies—drones hovered silently above the crowd, offering live feeds to command centers. They provided situational awareness without the intimidating presence of riot squads or surveillance towers. During active crime scenes, drones documented evidence and terrain from above without disturbing the physical ground or contaminating forensic traces.

Major police forces like the New York Police Department and Los Angeles Police Department established formal drone units, integrating them into citywide response networks. Coordination with fire departments, Homeland Security units, and emergency medical services became routine. UAVs helped locate hotspots in burning buildings, mapped storm-damaged neighborhoods, and monitored chemical spills—all from the air, without risking human responders.

But as police use grew more sophisticated, it quickly outpaced the legal framework meant to govern it. At the heart of the debate stood the Fourth

Amendment, the constitutional protection against unreasonable searches and seizures. Historically, courts had ruled that aerial observations—such as a helicopter flying over a backyard—did not constitute a search requiring a warrant. But drones introduced a new dimension to the conversation.

Unlike helicopters, drones could hover silently just above a backyard fence. They could peer into second-story windows, follow a suspect for hours without notice, and loiter unnoticed in tight or obscured spaces—capabilities far beyond the reach of traditional manned aircraft. They could be equipped with zoom lenses, infrared cameras, facial recognition software, and even AI-assisted behavior detection systems.

This raised urgent questions. At what point does persistent aerial monitoring become a form of surveillance requiring judicial oversight? Does the ability to hover and zoom transform public airspace into a tool of private intrusion? Should a UAV's advanced sensors change the threshold for what counts as a "search"?

Legal scholars, civil liberties organizations, and privacy advocates began sounding alarms. They warned of "surveillance creep"—the gradual normalization of technologies that erode expectations of privacy. What began as a tool for emergency response could, without clear guardrails, become an instrument of mass monitoring. The American Civil Liberties Union and Electronic Frontier Foundation both called for strict warrant requirements, transparency in drone use policies, and public disclosure of flight logs and data retention practices.

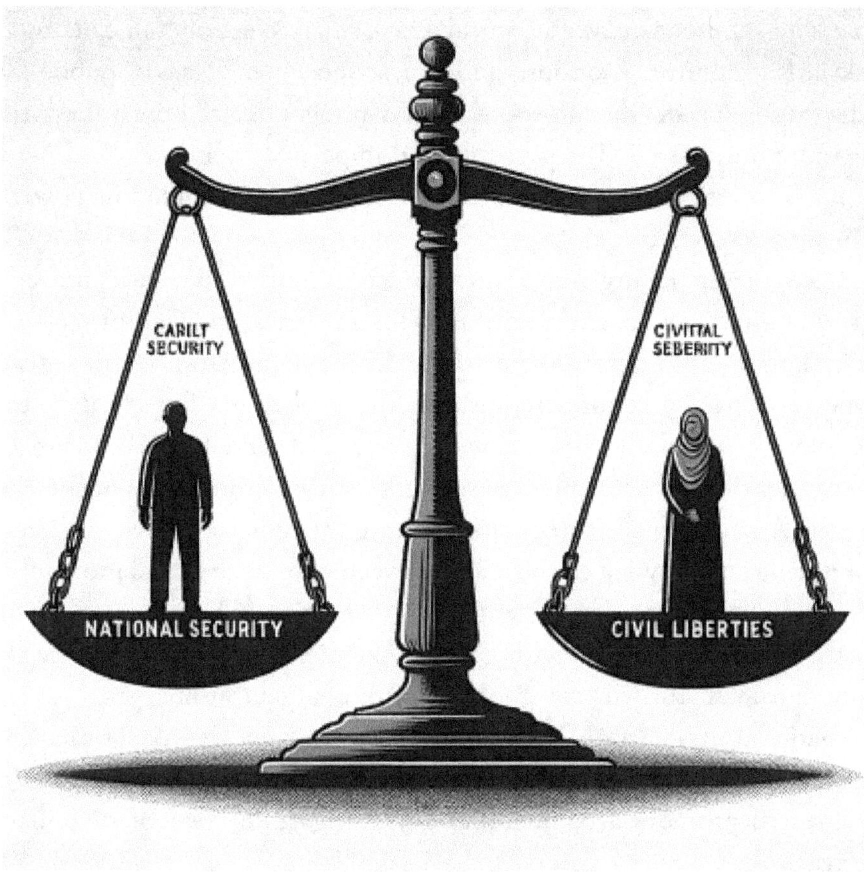

Yet regulation lagged. Federal guidelines were slow to materialize, and local ordinances varied widely. In some jurisdictions, police departments were required to obtain a warrant for any drone-based surveillance. In others, drones could be used freely, with little oversight or accountability. This patchwork approach left communities uncertain—and often unaware—of when, where, and how they were being watched.

In the space between innovation and legislation, drones quietly redefined the relationship between citizen and state. They offered new tools for safety and efficiency—but also raised new stakes for liberty and transparency. The future of drone-enabled policing would not be decided by technology alone, but by the legal and ethical frameworks society chose to build around it.

Civil liberties advocates have long warned that without clearly defined legal boundaries, the widespread use of drones in domestic surveillance could erode the expectation of personal privacy—especially in densely populated urban and suburban areas. In a world where small, silent aircraft can observe backyards, monitor rooftops, and track movement across public and semi-private spaces, the prospect of ever-present aerial observation risks becoming normalized. The psychological impact of "being watched" could change how citizens behave, gather, and speak—an invisible but powerful form of social control.

Recognizing the urgency of these concerns, several U.S. states moved to legislate boundaries around drone use. Florida, Texas, and California—each with large law enforcement agencies and histories of privacy advocacy—enacted statutes requiring judicial warrants for specific drone operations, particularly those involving prolonged surveillance or incursions into residential airspace. These laws sought to align drone deployments with Fourth Amendment protections, placing a judicial check on aerial monitoring and drawing clearer lines between public safety and intrusion.

In parallel, police departments began adopting transparency protocols to earn public trust. Faced with growing skepticism about "eyes in the sky," many agencies turned to guidance from organizations such as the International Association of Chiefs of Police and the American Civil Liberties Union. These groups recommended a suite of best practices designed to promote accountability and mitigate the fear of surveillance overreach.

Among the most effective strategies were the publication of formal drone-use policies. These documents defined the scope and limitations of drone missions, clearly identifying what drones could and could not be used for—routine patrols, yes; political protest monitoring, no. They also outlined strict retention and deletion rules for any video or imagery collected during operations.

Some departments implemented public accountability measures, delivering regular reports to city councils, civilian oversight boards, or independent review panels. These reports often included usage statistics, mission types, flight hours, and incident reviews—offering a layer of transparency rarely seen in traditional surveillance.

Community outreach became another key tool. In cities such as Austin, Sacramento, and Tampa, law enforcement agencies hosted open forums, invited residents to watch drone demonstrations, and fielded questions about the technology. These sessions aimed to demystify drone programs and make residents feel part of the conversation—not targets of it.

Data security was also a growing priority. With drones capable of recording high-resolution video, capturing license plate numbers, and geotagging movements, the need for strong data protection protocols became paramount. Many departments instituted encrypted storage, access logs, and automatic deletion after defined retention periods to prevent misuse or unauthorized surveillance.

In jurisdictions where these safeguards were implemented alongside meaningful public dialogue, drones began to shed their militarized image. Instead of being seen as symbols of state power, they were increasingly viewed as aerial tools for rescue, documentation, and disaster response.

Still, domestic drone programs face persistent challenges. One of the most insidious is mission creep. Systems acquired under the premise of supporting search-and-rescue operations or fire mapping are sometimes later used for low-altitude patrols or quiet monitoring of "high-crime" neighborhoods. The subtle shift from emergency aid to routine surveillance raises ethical questions about the long-term trajectory of UAV integration.

Airspace integration presents another dilemma. With drones increasingly common in populated areas, the Federal Aviation Administration continues to refine rules governing flight near airports, over crowds, and through complex urban environments. The risk of midair collisions, signal interference, or malfunctions grows with each additional UAV, placing pressure on cities to adopt unmanned traffic management systems.

Equity remains a third concern. Advocacy groups and civil rights leaders have noted that surveillance—particularly when unconstrained—often falls disproportionately on communities of color, immigrant neighborhoods, and politically active groups. The ability of drones to quietly monitor gatherings, movements, and associations raises fears of chilling effects on free speech and peaceful assembly.

As cities, courts, and communities wrestle with these dilemmas, the fundamental question persists: how to harness the value of drones without compromising the rights and freedoms they are meant to protect. In the absence of careful oversight, what begins as innovation could become intrusion. But with vigilance, dialogue, and restraint, drone technology may yet serve as a model for transparent, accountable integration of advanced tools into civic life.

As drone technology becomes cheaper, smarter, and more autonomous, the concerns surrounding their domestic use will only deepen. Advances in artificial intelligence, facial recognition, and persistent tracking capabilities may soon allow UAVs to follow individuals for hours or even days, identify them from afar, and cross-reference behavior against databases—all without human intervention. The potential for abuse is no longer speculative; it is technological inevitability waiting for policy to catch up.

The challenge ahead lies in balancing operational benefits with constitutional rights, ensuring that UAVs serve the public good without compromising democratic freedoms. Surveillance that goes unchecked, even when well-intentioned, risks normalizing a society in which anonymity is scarce and dissent is quietly recorded from above.

The integration of drones into homeland security and law enforcement represents one of the most significant domestic technology shifts since the arrival of radio dispatch or aerial photography. It is a transformation not just in how agencies operate—but in how they perceive, respond to, and manage public space. Drones now patrol international borders, scan forests for the missing, sweep rural highways after accidents, monitor parades, assess flood damage, and guide fire crews—all with a blend of speed, precision, and discretion that traditional tools cannot match.

And yet, their success depends on more than just capability. It depends on principled use. Public trust in aerial surveillance is not granted—it must be earned and maintained. If drones are to remain tools of safety rather than symbols of suspicion, agencies must commit to more than technological advancement. They must pair innovation with restraint, capability with consent, and surveillance with accountability.

Transparency will be critical. So will legal clarity. Just as law enforcement officers are held to procedural standards on the ground, UAV deployments must operate under rules that protect civil liberties in the air. In an age when nearly every American carries a device capable of recording, reporting, and broadcasting, the accountability burden does not rest with agencies alone—it will be shouldered by an informed and observant public.

The skies above the homeland are growing more crowded, not just with commercial aircraft and hobbyist drones, but with unblinking watchers that can record, analyze, and interpret human behavior in real time. The true test for society is not how high or how long these drones can fly—but how responsibly they are used, how openly they are governed, and how faithfully they reflect the values of a free and open republic.

Chapter 4: The Civilian Boom – From Hobby to Industry

What began as a military technology driven by Cold War exigencies and counterterrorism doctrine has, in the span of two decades, become a global phenomenon touching nearly every sector of civilian life. Drones—once limited to the high-security perimeters of military bases and intelligence outposts—have evolved into tools of agriculture, art, logistics, journalism, and recreation. They now hover over construction sites, glide through vineyards, trace the outlines of real estate listings, and capture cinematic footage once reserved for helicopters and cranes. Their ascent has spawned entire industries, sparked new regulatory regimes, and carved out cultural identities built around flight, speed, and aerial perspective.

The rise of civilian UAVs represents more than technological trickle-down; it is a full-blown renaissance in how people see, record, and interact with the world from above. From hobbyists piloting quadcopters in backyard fields to farmers using multispectral imaging to monitor crop health, drones have moved from secrecy to ubiquity—becoming as common in toolkits as tripods or tractors.

As of 2023, Chinese companies dominate this domain. DJI—Dà-Jiāng Innovations—stands unchallenged at the summit of the global civil drone market. The Shenzhen-based firm first crossed the 70% global share threshold in 2018, and has maintained that dominance ever since, becoming to drones what Apple is to smartphones or Boeing once was to commercial aircraft. DJI's product line spans a vast spectrum: from the compact Mavic series popular with filmmakers and travelers, to the enterprise-class Matrice and Agras platforms tailored for industrial and agricultural tasks. Their drones are celebrated for sleek design, reliable flight control systems, and unmatched camera stabilization—features that have set industry standards across continents.

No other manufacturer has come close. American firm 3D Robotics, once considered a potential rival, pivoted out of consumer drone production entirely. French company Parrot struggled to retain relevance after early success in the toy drone market. Even other Chinese competitors—Yuneec and Autel Robotics—have failed to surpass 5% of global market share. The result is an ecosystem where DJI products not only dominate the skies but also influence global expectations of what drones should do and how they should perform.

But dominance comes with scrutiny. Despite—or perhaps because of—its ubiquity, DJI has faced growing resistance from Western governments, particularly the United States. In 2020, the U.S. Department of the Interior grounded its entire DJI drone fleet, citing concerns over data security and potential exposure of sensitive location information to foreign servers. That same year, the Department of Justice issued a directive prohibiting the use of federal grant funds to purchase DJI or other foreign-made UAVs, citing national security risks tied to Chinese-manufactured hardware and software.

These actions were part of a broader trend: rising geopolitical suspicion over the use of Chinese technology in critical infrastructure and surveillance systems. Lawmakers and analysts expressed concern that flight data, video feeds, or telemetry could be accessed or exploited by Beijing through software backdoors or cloud storage agreements. Although DJI publicly denied any such vulnerabilities and offered data privacy modes for sensitive operations, the accusations triggered further investigations and fueled a push for domestic alternatives.

The regulatory headwinds did little to slow DJI's consumer popularity, but they signaled an emerging battleground over the future of civilian UAVs: one in which technical excellence would be matched—if not constrained—by questions of sovereignty, cybersecurity, and digital trust. The drone had become not only a commercial tool but a geopolitical flashpoint, hovering at the intersection of innovation and national interest.

By May of 2021, the United States had become a microcosm of the global drone revolution. The Federal Aviation Administration (FAA) reported more than 873,000 registered drones nationwide—42 percent for commercial use and 58 percent for recreation. What had once been a niche hobby or cutting-edge technology was now a permanent fixture in American skies. Consumer behavior underscored this transformation: in 2018 alone, sales of drones priced above $500 and $1,000 grew by over 33 percent year-over-year, revealing a clear appetite for more capable platforms equipped with advanced sensors, extended flight time, and autonomous features.

Unlike the military drone sector—dominated by a handful of aerospace giants in a few developed nations—the civil UAV market was decentralized and globally distributed. Startups, universities, and venture-backed firms from every continent began carving out niches in mapping, agriculture, inspection, logistics, and delivery. In the United States, drone companies pioneered solutions for last-mile parcel delivery, railroad inspection, and wind turbine maintenance. In India, drone innovators focused on agriculture, emergency medical logistics, and disaster response, with government-backed initiatives fueling rapid growth in rural UAV applications.

Academic institutions embraced the shift, building robust curricula around UAV engineering. Universities from California to Singapore offered dedicated degree tracks in aerospace robotics, computer vision, and unmanned systems. At the same time, a new industry emerged around drone education and licensing. Online training platforms, private flight academies, and certification prep courses proliferated to help commercial operators meet local aviation regulations—particularly under FAA's Part 107 rule, which governed licensed drone operations in U.S. airspace.

The civilian drone boom had far-reaching consequences. No longer a novelty, drones became embedded in the workflows of dozens of industries. They were deployed to assess wildfire damage in California, to map remote archaeological sites in Peru, and to deliver vaccines in Ghana. Their blend of precision, affordability, and versatility gave them an edge over helicopters, scaffolding, and even satellite imaging in many scenarios. Civilian drones found a home in:

Search and rescue, where thermal imaging drones combed forests and rubble for missing persons.

Law enforcement, which increasingly relied on drones to monitor large crowds, track fleeing suspects, or document crime scenes from above.

Medical logistics, with companies like Zipline deploying fixed-wing drones to deliver blood, vaccines, and emergency supplies across rural Rwanda—creating a model now emulated in Ghana, Nigeria, and parts of Southeast Asia.

Infrastructure inspection, where UAVs equipped with high-resolution cameras and LiDAR replaced climbers on power lines, inspectors on bridges, and surveyors on rail lines.

Precision agriculture, where multispectral imaging enabled farmers to monitor plant health, optimize irrigation, and apply inputs with surgical accuracy.

As drones became indispensable to civilian operators, they also caught the attention of armed forces and law enforcement agencies. Consumer UAVs—lightweight, maneuverable, and often modular—were increasingly co-opted for tactical missions. In 2018, the Israeli Defense Forces began integrating DJI quadcopters for forward reconnaissance and building surveillance. That same year, American police departments—from Los Angeles to rural Iowa—began fielding drones for crowd control, traffic accident analysis, and emergency response. In China's Xinjiang province, DJI drones were deployed as early as 2017 to monitor Uyghur communities, illustrating the dual-use nature of even the most commercially successful platforms.

Then came the most vivid and sobering demonstration of civilian drone power: the 2022 Russian invasion of Ukraine. On both sides of the battlefield, off-the-shelf drones were pressed into combat, modified to carry grenades, drop small bombs, or guide artillery strikes in real time. Hobbyists donated consumer drones by the thousands. Operators in trench networks and command bunkers alike used DJI models for reconnaissance, battlefield overwatch, and even psychological warfare—broadcasting drone footage of strikes to demoralize enemy forces.

The Ukrainian conflict became the first war in history to feature commercial drones at scale, transforming them from tools of convenience into weapons of survival. What had once been a cinematic accessory or an agricultural aid was now a frontline asset in 21st-century warfare.

At the epicenter of this global transformation stood one company: DJI. Founded in 2006 by Frank Wang, a graduate of the Hong Kong University of Science and Technology, DJI began as a student's engineering project. But through relentless innovation, intuitive design, and aggressive pricing, it grew into the most dominant force in the drone industry. By the late 2010s, DJI controlled more than 70 percent of the civilian drone market—a level of dominance virtually unrivaled in any other sector of consumer electronics.

Its drones flew on every continent, piloted by filmmakers, firefighters, police officers, farmers, and now soldiers. But this global success also brought scrutiny, regulation, and a new kind of warfare—one where the line between battlefield and backyard, between surveillance and service, grew increasingly thin.

By 2024, DJI controlled over 70 percent of the global drone market, producing models that set the standard across every category—from entry-level hobby drones to cinematic workhorses and industrial surveying platforms. The Mavic, Phantom, and Inspire series remained the most popular among consumers and professionals alike, celebrated for their stability, high-resolution imaging, compact design, and intelligent flight controls. Whether capturing cinematic aerials for Hollywood or mapping infrastructure for civil engineers, DJI drones became the default tools of choice.

Yet such market dominance did not come without controversy. U.S. lawmakers, defense officials, and cybersecurity analysts began raising alarms about data sovereignty and national security, citing the potential for telemetry and imagery gathered by DJI platforms to be accessed by Chinese intelligence services. These concerns intensified after the passage of China's 2017 National Intelligence Law, which mandates that Chinese firms cooperate with state security agencies upon request—casting doubt over the privacy of any data transmitted by Chinese-made devices, regardless of their intended use.

In response, the U.S. Department of the Interior grounded its entire fleet of DJI drones in 2020, halting environmental surveys, wildlife monitoring, and firefighting reconnaissance that had relied heavily on unmanned systems.

In 2021, the U.S. Treasury added DJI to an investment blacklist for its alleged involvement in surveillance activities within China, particularly the monitoring of ethnic minorities in the Xinjiang region. Bipartisan legislation soon followed, aiming to bar Chinese-made drones from federal use entirely. Though still under debate, such restrictions sparked deep concern across agriculture, journalism, emergency services, and infrastructure sectors—all of which had come to depend on DJI drones for affordable, reliable aerial tools unmatched by domestic alternatives.

While regulatory headwinds loomed, drone culture itself continued to evolve. One of the most captivating developments in recent years came from an unexpected arena: entertainment. Drone light shows—choreographed aerial performances using swarms of synchronized UAVs—began captivating audiences at festivals, national celebrations, and corporate events around the world. What began as a novelty soon gained traction as an alternative to traditional fireworks, offering not just visual spectacle but programmable precision.

These shows, often involving hundreds or even thousands of drones, combined GPS positioning with real-time flight control software to create animations, symbols, and dynamic forms in the night sky. They could form logos for product launches, recreate national flags during holidays, or simulate fireworks with flawless timing and zero risk of combustion. Cities from Tokyo to Las Vegas embraced the format, especially during periods of fire risk or environmental concern.

The appeal was multifaceted. Unlike fireworks, drone shows posed no risk of fires, toxic smoke, or falling debris. Their silence minimized impact on wildlife and urban noise pollution. Moreover, the flexibility of programmable routines opened new frontiers in visual storytelling—from space-themed sequences celebrating lunar landings to intricate cultural motifs during Lunar New Year festivals. The result was not only safer and cleaner than explosives, but more expressive, interactive, and sustainable.

As drone hardware became smarter and more compact, the boundary between utility and art blurred. No longer confined to the battlefield or job site, drones had become canvases of creativity—pixelated points of light suspended in orchestrated flight, transforming public skies into theaters of innovation.

Major companies such as Intel, SKYMAGIC, and Verge Aero have pioneered large-scale performances at global events including the Olympics, World Expos, and Super Bowl halftime shows. These spectacles, often involving thousands of drones operating in tight synchrony, transformed night skies into living murals of light and motion. Some hybrid shows now integrate drones carrying pyrotechnics, creating novel combinations of animation, color, and sound that blur the line between art installation and aerial technology. The result is a sensory experience unlike anything achievable with traditional fireworks—dynamic, programmable, and endlessly adaptable.

Meanwhile, a different form of drone performance has surged from grassroots innovation to global spectacle: drone racing. What began as a hobbyist pastime evolved into a fast-growing subculture and professional sport defined by speed, skill, and immersive technology. Pilots fly small, high-thrust quadcopters through complex obstacle courses at speeds exceeding 100 miles per hour, often using first-person view (FPV) headsets that stream real-time video from onboard cameras. This cockpit-like perspective creates an exhilarating, high-intensity experience, not only for pilots but for spectators watching digitally enhanced feeds.

Racing formats vary widely—from solo time trials to head-to-head eliminations and team-based relays. Organized leagues such as the Drone Racing League (DRL) and MultiGP have brought this high-speed sport to mainstream audiences through televised broadcasts, esports tie-ins, and even augmented reality overlays. Competitions showcase not only the piloting reflexes of competitors, but also the engineering ingenuity behind each craft—where weight, thrust, firmware optimization, and custom frame design play crucial roles in determining victory. Drone racing has become a digital-era motorsport, drawing together a community of makers, programmers, and adrenaline junkies into a vibrant new aviation frontier.

Drones have also transformed creative expression, particularly in the realm of aerial photography and filmmaking. What once required helicopters, cranes, or expensive camera rigs is now achievable with a consumer UAV and a bit of practice. Today's drones, even at entry-level price points, come equipped with 4K cameras, gimbal stabilization, autonomous tracking, and pre-programmed cinematic flight paths. For filmmakers, journalists, and artists, this has unlocked new ways to capture landscapes, follow action, or tell stories from elevated and mobile perspectives. The democratization of aerial imaging has reshaped visual storytelling—from wedding videos and real estate marketing to independent documentaries and breaking news coverage in war zones and disaster areas.

For more complex shoots, professional aerial cinematography drones like the AERIGON require coordinated teamwork. Typically, one operator pilots the drone, while a second controls the camera, managing focus, aperture, and composition in real time. These dual-operator systems are now standard in high-end film and television production, with drones capturing dramatic sequences for Hollywood blockbusters, wildlife documentaries, and National Geographic expeditions in environments too risky or remote for helicopters. They have opened visual frontiers, enabling directors to choreograph shots once limited by cost or logistics—swooping through canyons, tracing mountaintops, or gliding silently across arctic tundras.

Drones have also become indispensable for filming in hazardous zones. Their ability to reach collapsed buildings, flooded neighborhoods, unstable cliffs, or protected wildlife habitats allows journalists, scientists, and first responders to gather imagery without endangering human life. These capabilities have spawned new disciplines: drone journalism, which offers fast, on-the-ground coverage in active war zones or protest movements; environmental videography, documenting glacial melt, deforestation, or coral bleaching from above; and virtual tourism, where drone footage delivers immersive experiences of landmarks and ecosystems to global audiences without leaving home.

In industry after industry, drones are proving to be more than flying cameras—they are workhorses capable of real-time monitoring, autonomous flight, and efficient data capture. In agriculture, they are transforming how food is grown and managed. Precision agriculture—an approach that uses data to guide farming decisions—is now reliant on UAVs to optimize both yield and resource use. Drones equipped with multispectral sensors and thermal cameras can identify early signs of pest infestation, crop stress, or fungal infections long before they are visible to the human eye. They fly over fields collecting infrared and visible light data, building detailed crop health maps that guide targeted irrigation or pesticide deployment.

Advanced UAVs in farming now carry tanks capable of delivering pesticides, herbicides, and fertilizers directly to affected zones. These autonomous systems follow GPS-guided routes across hundreds of acres, working day or night, and adapting to terrain with onboard obstacle sensors. Companies such as XAG and Rotor Technologies have developed drone sprayers with payload capacities exceeding 100 gallons, effectively replacing light aircraft in many spraying operations. The result is reduced chemical waste, lower labor demands, and improved environmental stewardship.

As global agriculture faces increasing pressure from climate change, water scarcity, and workforce shortages, drones are emerging as a vital part of the solution. The agricultural UAV sector is expected to grow from $3.6 billion in 2024 to $5.7 billion by 2030. In this next chapter of farming, drones do not merely monitor—they manage, respond, and optimize, helping feed a world in flux.

In the film and television industry, drones have shattered creative limits. What once required helicopters, cranes, and elaborate rigging can now be accomplished with a backpack-sized quadcopter operated by a single technician. High-end systems like DJI's Inspire 3 offer professional-grade features once limited to aerial cinematography crews with six-figure budgets. These drones come equipped with omnidirectional obstacle sensing, 6K or 8K RAW video capture, dual-operator controls for independent camera and flight operation, and real-time GPS stabilization that enables precision tracking in even the most dynamic environments.

Blockbuster films, automotive commercials, and wildlife documentaries increasingly rely on drones to deliver sweeping landscape shots, overhead action sequences, and smooth tracking movements that once required extensive planning and risk. This technology has democratized access to cinematic-quality aerial footage. Independent filmmakers and content creators now work with tools that rival what major studios once reserved for helicopter-mounted rigs—transforming not only budgets, but visual storytelling itself.

News organizations have also embraced drones as essential tools for frontline reporting. Whether covering natural disasters, political protests, or active war zones, UAVs allow journalists to capture real-time aerial perspectives without placing camera crews in harm's way. During hurricanes, wildfires, and floods, drones provide views of structural damage, evacuation routes, and environmental impact. In conflict zones or areas under tight military control, they offer a form of visual documentation often beyond the reach of conventional reporting.

Networks like the BBC, CNN, and Al Jazeera have invested in dedicated drone teams, turning UAVs into standard gear for modern journalism. Beyond event coverage, drones have enhanced investigative reporting. They have tracked illegal deforestation in the Amazon, filmed glacial retreat in the Arctic, and exposed unauthorized construction or encampments in disputed territories—expanding the scope of public transparency.

In logistics, drones are reshaping the "last mile"—the final and often most challenging segment of delivery. For years, this phase has represented a bottleneck in both cost and efficiency, especially in rural regions, high-density cities, or during crisis scenarios. Companies such as Amazon Prime Air, Alphabet's Wing, and Zipline have pioneered unmanned delivery trials, using fixed-wing and multirotor UAVs to transport medical supplies, pharmaceuticals, diagnostic kits, and consumer goods directly to customers or field hospitals.

These programs have demonstrated remarkable potential. In Rwanda and Ghana, Zipline's drone network now delivers blood and vaccines to remote clinics in under 30 minutes. In the United States and Australia, test programs have brought household items and healthcare products to suburban households, bypassing traffic and supply-chain bottlenecks. During the COVID-19 pandemic, drones played a crucial role in delivering test kits and protective gear to lockdown zones and quarantine sites.

Still, large-scale deployment of delivery drones faces significant hurdles. Integration with manned aviation remains a regulatory challenge, especially in crowded urban airspace. Safety concerns—including mid-air collisions, property damage, or privacy violations—must be addressed through rigorous testing and oversight. Noise pollution and public acceptance are further concerns, particularly when operations extend into residential neighborhoods.

Despite these challenges, the trend is clear: UAVs are becoming indispensable tools in media, logistics, and public health. The question is not whether drones will reshape these industries—but how quickly the regulatory and technological infrastructure can adapt to meet the demand.

Despite these challenges, pilot programs in Australia, Ghana, Rwanda, and Virginia have proven that drone delivery is not only viable, but often optimal—especially in regions with limited infrastructure or difficult terrain. In such contexts, UAVs bypass logistical bottlenecks, providing rapid and targeted access to essential supplies.

Parallel to commercial and industrial applications, drone flying has blossomed into a vibrant recreational and semi-professional culture. What began as a niche hobby has grown into a global community of enthusiasts who use drones for aerial photography, cinematic storytelling, competitive racing, and artistic experimentation. First-person view (FPV) drones, in particular, have opened new possibilities in immersive flight experiences, enabling pilots to navigate complex 3D environments through onboard cameras and virtual goggles.

Online platforms such as YouTube and TikTok have become central to this subculture. Influencers like Ken Heron, Billy Kyle, and Captain Drone have attracted substantial followings by posting unboxings, flight tutorials, gear reviews, and visually stunning footage. These creators not only educate but also inspire, contributing to a cultural shift that has elevated drones from tools to toys to artistic instruments.

Drone racing, once limited to garages and open fields, has evolved into a high-stakes global sport. The Drone Racing League (DRL) hosts televised competitions featuring high-speed drones outfitted with LED lighting systems, zipping through elaborate courses at speeds exceeding 90 miles per hour. With its fusion of real-world engineering and esports-style presentation, drone racing appeals to both STEM-minded audiences and adrenaline-seeking fans, pushing the limits of piloting skill and hardware customization.

To fly drones commercially in the United States, operators must be certified under the Federal Aviation Administration's Part 107 regulation—a framework established in 2016 to formalize the growing role of UAVs in business and public operations. Under Part 107, drone pilots must be at least 16 years old, pass the FAA's Aeronautical Knowledge Test, and undergo Transportation Security Administration (TSA) background screening.

The certification process ensures that operators understand essential aviation principles, including airspace classifications, weather patterns, and emergency protocols. Pilots are trained in the rules governing night operations, the use of waivers, and the strict avoidance of restricted areas such as airports, government facilities, and infrastructure deemed critical to national security.

The FAA also maintains the Low Altitude Authorization and Notification Capability system—LAANC—a digital platform that enables commercial drone operators to request and receive near-real-time flight authorization in controlled airspace. By automating and accelerating the approval process, LAANC has become an indispensable tool for businesses operating near airports or in densely regulated urban environments. It reflects a broader effort to modernize airspace management as UAV traffic increases.

The civilian boom in drone use reflects more than a passing technological trend—it signals a profound reimagining of how humanity engages with the sky. In previous generations, aerial perspectives required helicopters, broadcast vans, or military-grade surveillance systems. Today, a compact drone and a smartphone can deliver cinematic footage, inspect power lines, or deliver medical supplies across mountain passes. Drones have democratized flight itself, distributing capabilities once confined to government and industry into the hands of farmers, artists, engineers, and entrepreneurs.

But as their presence multiplies and their intelligence deepens, drones are reshaping the very fabric of the airspace they inhabit. No longer merely flown—they are sensed, mapped, networked, and increasingly autonomous. They form part of a growing aerial ecosystem in which flight is not only about altitude and velocity, but about bandwidth, data fusion, and algorithmic precision.

Managing this invisible, digitized sky will require more than regulatory enforcement. It will demand vision: an integration of innovation with ethical foresight. It must preserve the freedom to explore while protecting the right to privacy, ensure safety without stifling creativity, and maintain equitable access in a domain once ruled by exclusion.

The story of drones—once whispered in classified briefings and coded in military acronyms—now belongs equally to the orchard, the newsroom, and the neighborhood park. The sky is no longer the limit. It is the new commons.

Chapter 5: Brains in the Sky – The Technology Behind Autonomous and Swarming Drones

The story of drones is no longer defined solely by wingspan, payload, or endurance—it is increasingly a story of intelligence. In recent years, the most consequential breakthroughs in unmanned aerial systems have come not from new airframes, but from what lies within: artificial intelligence, real-time data processing, and autonomous coordination.

From individual drones capable of making split-second decisions without human input to swarms that mimic the adaptive behavior of insect colonies, this "brains in the sky" revolution is transforming UAVs from remotely piloted vehicles into independent agents. These platforms now execute complex, collaborative missions in dynamic, unpredictable environments. The shift brings not only immense tactical and logistical advantages—but also a host of profound ethical and strategic dilemmas.

Artificial Intelligence (AI) has emerged as the cornerstone of this evolution. By allowing UAVs to interpret their surroundings and act upon real-time data, AI transforms aircraft into sensing, reasoning machines. Modern drones are no longer limited to following pre-programmed waypoints; they now adjust flight paths, identify targets, and assess risk dynamically—all while in motion.

Today's intelligent UAVs can:

· Detect and avoid obstacles in three-dimensional space

· Identify individuals or objects using facial and pattern recognition

· Track moving targets with accuracy exceeding 98%

· Recalculate optimal routes in response to terrain, threats, or signal jamming

Companies like Shield AI have pioneered this capability through systems like Hivemind, an AI autonomy stack enabling drones to operate effectively in GPS- and communications-denied environments. These drones do more than fly; they explore, adapt, and complete their missions using onboard logic and sensor fusion. In a collapsed building, a jungle canopy, or a contested combat zone, such capabilities can mean the difference between success and failure.

The leap from operator-guided to self-directed UAVs has been fueled by several converging technologies:

· Edge computing, which empowers drones to process and analyze sensor data in real time without relying on distant servers

· Machine learning, which enables platforms to "learn" from experience and improve over successive missions

· Sensor miniaturization, allowing UAVs to carry high-resolution cameras, radar, LiDAR, and environmental sensors within a compact frame

New platforms such as the MQ-35 V-BAT, a vertical takeoff and landing (VTOL) system, demonstrate how autonomy is reshaping battlefield and emergency response operations. Designed to function in contested electronic environments, the V-BAT can complete reconnaissance and surveillance tasks using onboard algorithms alone—without satellite guidance or pilot commands.

Across civilian and military sectors, intelligent drones are now autonomously performing: Large-area mapping and persistent surveillance

· Infrastructure inspection, including real-time defect identification using AI vision .· Time-critical delivery of medical supplies or food in disaster zones and hard-to-reach communities

As intelligence becomes the defining characteristic of next-generation UAVs, the boundaries between aircraft and autonomous agents begin to blur. These systems are no longer just tools; they are decision-makers. The implications—legal, moral, and operational—will shape the future not only of aviation but of warfare, logistics, and civil governance itself.

These operations, once impossible without pilot input, are now increasingly routine—and growing more complex with each technological advance.

Perhaps the most transformative frontier in modern drone development is the rise of autonomous swarms. Inspired by the collective behaviors of bees, birds, and other natural groups, drone swarms represent a new paradigm in unmanned systems—cooperative, distributed, and remarkably resilient. Rather than being directed by a single ground operator or centralized controller, each drone within a swarm operates semi-independently, constantly reacting to its environment and the positioning of its neighboring drones. Together, they exhibit emergent behavior—patterns and intelligence arising not from top-down commands, but from simple, localized interactions.

This decentralized model offers key advantages. If one drone fails or is destroyed, the others adjust and continue the mission with no loss of cohesion. The system is inherently scalable; dozens or even hundreds of drones can be added without overloading command structures. And it is extraordinarily adaptable, capable of reorganizing itself in real time in response to terrain, weather, or enemy actions.

The concept moved from theory to reality in 2016, when the U.S. Department of Defense conducted a groundbreaking demonstration under the Perdix program. Over the test ranges of China Lake, California, three F/A-18 fighter jets released more than one hundred micro-drones, which immediately began operating as a coordinated swarm. They flew in formation, avoided obstacles, and mapped the area—entirely autonomously, without direct operator input. It was the first large-scale test of swarm intelligence in the field, and it worked.

Since then, similar efforts have taken root around the world. Hungarian researchers have fielded swarms of over a hundred drones for environmental mapping, pushing the limits of coordination and real-time data collection. Chinese defense contractors have showcased synchronized launches of swarm drones from trucks, boats, and even wearable packs, clearly signaling battlefield intentions. Russia and India, likewise, have invested heavily in loitering munitions that swarm and saturate defenses with coordinated attacks. What once seemed a distant possibility has become a race among military powers.

These capabilities depend not just on clever algorithms, but on a dense web of enabling technologies. Navigation remains foundational. Most drones today rely on satellite-based positioning systems such as GPS, GLONASS, or Galileo, enabling them to determine their position and follow predefined paths. For applications requiring pinpoint precision—agricultural spraying, construction mapping, or industrial surveying—many now use real-time kinematic positioning, refining their location to within centimeters.

Equally vital is the integration of sensors that allow drones to perceive and interpret their surroundings. Inertial measurement units provide information about orientation and movement, magnetometers function as onboard compasses, and altimeters and barometers keep track of altitude. More advanced drones use ultrasonic rangefinders and infrared sensors to detect nearby obstacles. High-end models employ LiDAR or photogrammetry systems to build three-dimensional maps of terrain and structures.

Taken together, these technologies enable drones not only to fly safely and efficiently, but to understand their environment—and each other. In swarms, this shared perception becomes a networked intelligence, one capable of mapping terrain, identifying targets, relaying communications, and responding to mission changes in real time.

The age of the autonomous drone is not coming—it has arrived. And in the shape of swarms, we are beginning to see a new kind of machine behavior: one that is decentralized, adaptive, and potentially far more powerful than any single drone alone.

Machine Vision

Computer vision has become the eyes of modern drones, converting streams of video into actionable intelligence in real time. These systems, powered by onboard neural networks, now enable UAVs to identify moving targets, detect cracks or corrosion in bridges and buildings, recognize hand gestures or QR codes, and track individuals even as they move through crowds or dense terrain. Such capabilities are no longer experimental—they are essential in a growing list of civilian and military applications, from inspecting power lines and monitoring endangered wildlife to surveilling conflict zones and aiding search-and-rescue operations.

At the center of this intelligence is the flight controller—effectively the drone's brain. What began as crude analog circuits and relay switches in the mid-20th century has evolved into compact, aviation-grade computing platforms known as autopilots or flight control boards. These systems fuse data from accelerometers, gyroscopes, magnetometers, barometers, and GPS receivers to calculate orientation, velocity, and position. In high-performance

UAVs, these sensors provide ten or more degrees of freedom (DOF), enabling drones to navigate autonomously through highly dynamic environments. Dual-processor setups provide redundancy in critical systems, and real-time decision loops maintain stability and responsiveness.

In 2024, the European Union Aviation Safety Agency issued its first flight controller certification under the new ETSO-C198 standards, a landmark moment that recognized the Embention Veronte autopilot as a system fit for integration into regulated airspace. This signaled the maturity of drone avionics and marked an inflection point in their adoption for large-scale logistics, urban air mobility, and beyond-visual-line-of-sight (BVLOS) operations.

Open-source platforms like PX4 and ArduPilot continue to fuel innovation at the hobbyist and commercial levels, supporting everything from crop-spraying hexacopters to autonomous delivery drones. Meanwhile, high-end systems like the Veronte Autopilot 1x are pushing the boundaries in certified aviation environments. These controllers do more than fly—they interpret missions, manage health diagnostics, interface with payloads, and increasingly plug into AI modules that allow UAVs to learn from experience and make decisions on the fly.

But with this rise in autonomy comes a deepening dilemma. The prospect of lethal autonomous weapons systems—drones capable of selecting and engaging targets without human oversight—has triggered global alarm. If a drone commits a war crime or misidentifies a target, who bears responsibility? The programmer? The commander? The state? Or the algorithm?

Worries extend beyond accountability. Autonomous swarms could accelerate conflict in ways that outpace human decision-making. Their proliferation threatens to lower the threshold for armed violence, empowering rogue states and non-state actors with scalable killing tools that require no infrastructure or command chain. And all of this is occurring in a legal vacuum: the Geneva Conventions never imagined software-guided weapons operating with no human in the loop.

Calls for regulation have grown louder. The United Nations, Human Rights Watch, and dozens of civil society organizations have urged states to prohibit or strictly regulate fully autonomous lethal drones. Yet while some nations advocate for bans, others see such systems as vital to their national defense futures. As technology races ahead of diplomacy, the world faces a critical test—not just of engineering, but of ethics, law, and leadership in the age of intelligent machines.

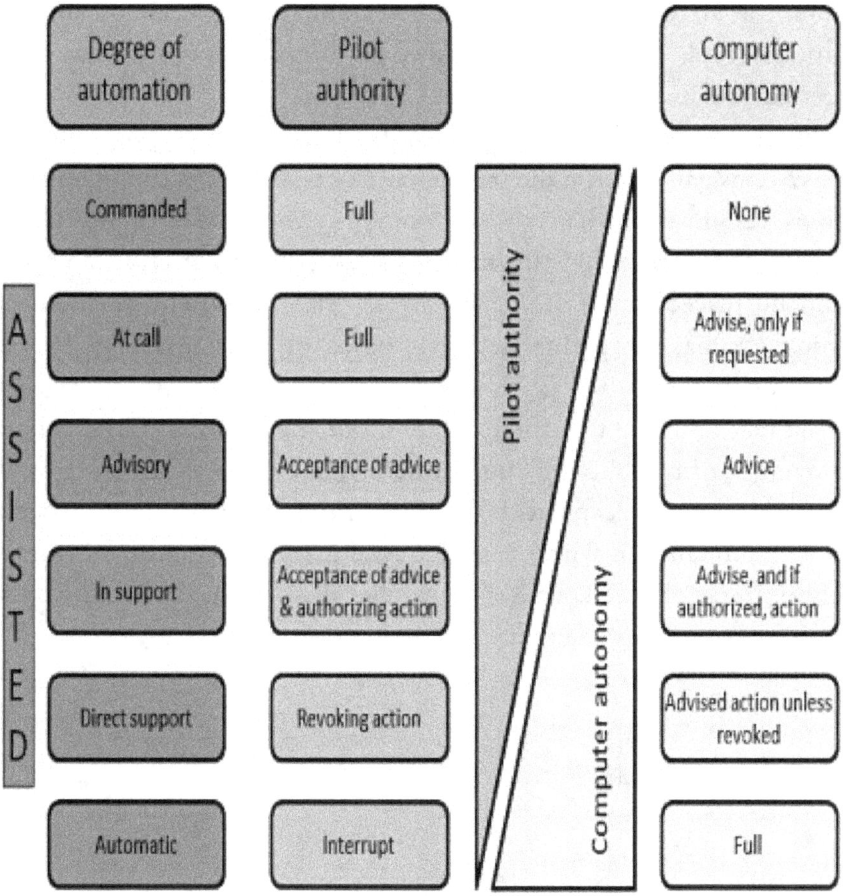

The shift from remotely piloted drones to fully autonomous aerial agents signals more than just a technological leap—it marks the dawn of a fundamentally new era in aviation. No longer tethered to the joystick or reliant on continuous operator input, these machines now possess the capacity to interpret their environments, make decisions, and coordinate with one another in real time. In many scenarios, they act not just as extensions of human will, but as semi-independent actors operating faster and more efficiently than human reflexes allow.

This transformation offers breathtaking potential. Autonomous drones can deliver medical supplies deep into disaster zones inaccessible by road or helicopter. They can penetrate smoke-filled skies to locate wildfire hotspots and track their spread. In cities, they can scan infrastructure, map terrain, or monitor air quality with a precision and persistence unmatched by traditional tools. Their promise is vast—measured in lives saved, time reclaimed, and insights gained.

Yet with such power comes a complex web of new responsibilities. As these systems gain agency, the traditional boundaries of accountability blur. Who is responsible when an autonomous drone makes a mistake—an engineer, an operator, a policymaker, or the machine itself? What limits should be placed on drones that can surveil indefinitely, strike without warning, or make targeting decisions based on pattern recognition and algorithms alone?

The real test is not in the machines, but in ourselves. The technology is moving swiftly—but will our ethical frameworks, legal systems, and international norms keep pace? The skies of tomorrow will be crowded not only with aircraft, but with questions. And if we are to meet this moment wisely, we must chart a course that matches the brilliance of our inventions with the depth of our judgment.

Chapter 6: The Ethics and Risks of an Unmanned World

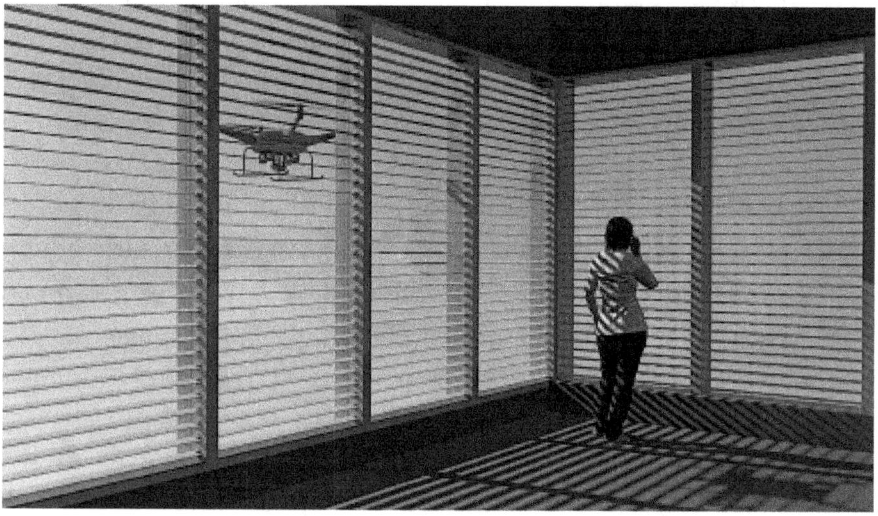

The rise of drones has reshaped not only battlefields and industries but also the moral and legal frameworks that underpin modern society. As unmanned systems grow smarter, faster, and more autonomous, their impact deepens—challenging traditional boundaries of privacy, security, accountability, and international law.

Drones equipped with high-resolution cameras, infrared sensors, and long-endurance batteries can now loiter silently over homes, public parks, or city intersections—capturing data without the subject's awareness, let alone consent. Law enforcement agencies, journalists, private companies, and even hobbyists have used drones to observe individuals in both public and quasi-private spaces. This has led to a growing number of legal gray zones and rising public backlash. While courts have often ruled that aerial observation from public airspace does not always constitute a "search" under the Fourth Amendment, the transformative capabilities of modern drones have changed the calculus. The ability to hover quietly, zoom in from hundreds of feet away, and track individuals in real time pushes existing legal standards to their limits.

This is no longer a hypothetical scenario. In Baltimore, a controversial aerial surveillance program operated by police involved drones recording thousands of hours of citywide footage. When it was revealed that this data was being used to retroactively track individuals without warrants, public outcry followed. Critics argued that the program constituted mass surveillance by stealth—quiet, constant, and largely invisible to those being monitored. Such capabilities, if left unchecked, could usher in an era of persistent, algorithmic observation where civil liberties are eroded not by force, but by optics, data storage, and software.

The accessibility of drone technology has also introduced a host of new security vulnerabilities. Consumer-grade UAVs, available for a few hundred dollars, have been repurposed for illicit activities around the world. Inmates have received contraband—including drugs, phones, and weapons—delivered via drone drops into prison yards. Corporate espionage has taken to the skies, with drones conducting unauthorized surveillance of industrial sites, competitor facilities, and sensitive infrastructure. In multiple cases, drone incursions near commercial airports have led to grounded flights, panicked air traffic controllers, and even partial shutdowns of international airports.

The threat is not confined to amateurs or criminals. Non-state actors and terrorist groups have adopted drones for reconnaissance and as makeshift weapons. In Iraq and Syria, ISIS fighters equipped commercial drones with improvised explosive devices, transforming them into low-cost suicide bombers. In Venezuela, an attempted assassination of President Nicolás Maduro involved drones rigged with explosives flown over a public event. These examples underscore a stark reality: drones have democratized not only surveillance, but also the means of attack.

As the line between civilian and combatant blurs in the unmanned age, so too does the boundary between innovation and intrusion. The legal, ethical, and societal frameworks developed for traditional aviation, telecommunications, and personal privacy are no longer sufficient. A new doctrine must emerge—one that recognizes the extraordinary capabilities of modern UAVs while enforcing meaningful limits on their use

By 2024, the scale of the drone challenge had become impossible to ignore. Reports indicated that two-thirds of U.S. airports had experienced near-miss incidents involving unauthorized drones, some flying within mere yards of jet engines during takeoff or landing. The FAA continued its efforts to develop and deploy countermeasures—ranging from drone detection networks to radio-frequency jamming systems—but enforcement remained a daunting and incomplete task. As drone numbers soared, the airspace above America's cities and runways became increasingly congested, unpredictable, and vulnerable.

More troubling still was the deliberate weaponization of drones by terrorist organizations and transnational criminal groups. Their low cost, accessibility, and near-silent operation made them ideal tools for asymmetrical warfare. Drug cartels used drones to drop explosives over police convoys in Mexico, while insurgent groups in the Middle East attached grenades or IEDs to off-the-shelf quadcopters to strike U.S. and allied forces. In one of the most dramatic incidents to date, an explosive-laden drone targeted Venezuelan President Nicolás Maduro during a televised public event in 2018. Though the attack failed, it vividly illustrated the expanding reach of unmanned threats—even to heads of state.

Yet perhaps the most chilling prospect in drone development is the emergence of lethal autonomous weapons systems, or LAWS. These are machines programmed to select and engage human targets without human intervention—driven by advances in artificial intelligence, machine learning, and sensor integration. The U.S. Department of Defense has maintained a formal policy requiring "meaningful human control" over any use of lethal force. But that line is becoming increasingly difficult to enforce as autonomy grows more capable and rivals like China and Russia accelerate their own autonomous weapons programs. From facial recognition and behavioral pattern analysis to AI-driven threat prioritization, the components of LAWS are already being tested, if not yet fully deployed.

Ethicists, human rights advocates, and legal scholars have raised alarms about a future where machines may hold the power to kill without accountability. Delegating lethal decisions to software, they argue, violates fundamental principles of human dignity, proportionality, and the laws of war. What if a targeting algorithm misidentifies a child carrying a shovel as an enemy combatant? What happens if adversaries spoof facial recognition systems or manipulate the AI's decision-making matrix? These are not speculative concerns—they are technical vulnerabilities already identified in experimental systems.

In response, the United Nations and advocacy coalitions like the Campaign to Stop Killer Robots have called for a binding international treaty to prohibit the development and use of fully autonomous weapons. Yet momentum has stalled. The world's leading military powers have refused to commit, wary of constraining their capabilities in a new AI-driven arms race that few believe can be avoided—and none want to lose.

Even in the absence of full autonomy, the ethical distance between operator and target has grown dramatically in the drone age. Remote pilots, sitting in air-conditioned trailers thousands of miles from the battlefield, watch grainy video feeds for hours—observing the habits, movements, and lives of suspected militants before executing strikes. At the end of their shift, they drive home through quiet neighborhoods, returning to family dinners or children's soccer games. This duality—lethal warfare by day, domestic normalcy by night—has created a unique psychological burden. While physical risk is eliminated, moral injury and emotional detachment linger.

Some fear that this clinical model of warfare desensitizes both operators and policymakers. When no pilots are placed in harm's way, the political cost of authorizing strikes diminishes. Targeted killing becomes easier to justify, simpler to execute, and harder to question. In such an environment, war itself risks becoming not a last resort—but an algorithmically administered norm.

Would you like to continue into a closing section that reflects on the future trajectory of drones and the societal choices ahead?

The August 2021 drone strike in Kabul, which tragically claimed the lives of 10 civilians—including seven children—was a grim reminder that even the most precise weapons are only as reliable as the intelligence behind them. Intended to prevent a terrorist attack during the chaotic withdrawal from Afghanistan, the strike instead targeted an aid worker mistaken for a militant. The Pentagon later admitted its error, but the damage—to lives, trust, and the credibility of drone operations—was already done. It underscored a core truth of modern warfare: that precision is not infallibility, and that technological prowess cannot compensate for flawed assumptions or incomplete data.

As drones grow more sophisticated and connected, they also become more vulnerable. Every enhancement in autonomy or communication brings with it a new surface for cyberattack. Hackers can spoof GPS coordinates, misleading a drone about its true location and sending it astray. Unsecured command links invite hijacking—allowing an adversary to assume control mid-flight. Surveillance drones, if not properly encrypted, risk transmitting sensitive imagery or telemetry over exposed networks. These are not theoretical dangers. In 2011, Iranian forces famously captured an intact U.S. RQ-170 Sentinel by reportedly spoofing its GPS, coaxing it into landing on Iranian soil. The incident provided Tehran with a trove of advanced technology and offered the world a case study in the fragility of even the most advanced unmanned systems.

The legal landscape surrounding drone warfare remains equally fraught. Cross-border strikes in nations like Pakistan, Yemen, Somalia, Libya, and Iraq routinely test the boundaries of state sovereignty and international law. While the United States has often cited Article 51 of the U.N. Charter to justify such actions as self-defense against non-state actors, many of the targeted countries and independent observers view these strikes as violations of territorial integrity. The opaque nature of target selection, coupled with the lack of judicial oversight and post-strike accountability, has fueled criticism from human rights organizations and legal scholars alike. Without an agreed-upon global framework for drone use—especially in extraterritorial operations—the risk persists that drone warfare will normalize a world of perpetual, unregulated assassination.

This ambiguity is further exacerbated by the increasing reliance on private contractors, outsourced intelligence, and proxy forces. When drone missions are run by non-state actors or third-party operatives, attribution becomes difficult. Who bears responsibility when a strike goes wrong—the operator, the client, the software designer? In an era of distributed warfare, the answer is often murky.

Economically, drones are a double-edged sword. In agriculture, they can monitor crops, detect blight, and distribute fertilizer with unprecedented efficiency. But this precision can also reduce the need for seasonal labor, displacing workers in already precarious economies. In journalism and filmmaking, drones provide breathtaking new angles—but they also raise new ethical questions about voyeurism, consent, and the line between reporting and intrusion. In logistics, autonomous drones offer faster, cheaper delivery of goods and medicine—yet they threaten jobs in warehousing, transportation, and courier services.

The democratization of drone technology is uneven. Wealthier nations, corporations, and institutions can afford fleets of advanced UAVs for public safety, disaster response, or infrastructure monitoring. Meanwhile, poorer communities may lack access to even basic systems, widening the gap in preparedness, communication, and opportunity. The drone revolution, like all technological shifts, brings great promise—but without equitable policies, it may deepen the very divides it could help solve.

As drone use accelerates across every sector of society, so too must the systems that guard against their misuse. Counter-unmanned aerial vehicle (C-UAV) technologies are now a rapidly growing field, driven by the need to secure airports, military installations, government facilities, and urban airspace from both hostile and negligent drone activity. These defense systems vary widely in complexity and application—from radio frequency jammers that disrupt control signals, to interceptor drones that launch nets mid-air, to high-powered directed energy weapons capable of disabling UAVs with microwaves or lasers. Some use artificial intelligence to analyze flight paths, distinguish drone types by acoustic signatures, and predict behavior in real time. Companies like Dedrone, Fortem Technologies, and Anduril Industries are pioneering these tools. Yet no single solution is foolproof—especially against swarms, stealthy low-flying microdrones, or fully autonomous systems operating outside traditional detection protocols.

The unmanned future brings with it the imperative for a new ethical and legal architecture—one that doesn't lag behind the machines it's meant to govern. Global stakeholders must craft robust frameworks to ensure these technologies serve humanity rather than undermine it. Key proposals include internationally binding treaties that ban or strictly regulate lethal autonomous drones, legal standards assigning responsibility to developers and commanders, and transparency mandates that compel agencies to disclose the scope, use, and oversight mechanisms behind both domestic surveillance and military strikes. Public involvement is critical; policies concerning privacy, accountability, and civil liberties must be shaped not in secrecy, but through democratic dialogue. Drone systems themselves must be designed with ethical principles at their core—embedding safety checks, fail-safes, and usage constraints into their very code and architecture.

In many ways, the emergence of autonomous drones forces us to revisit fundamental questions: Who is accountable when machines make life-or-death decisions? How do we protect personal privacy in an age of algorithmic surveillance? What safeguards prevent the automation of violence or oppression? These are not questions for tomorrow—they are questions for now.

The trajectory of drones will not be shaped solely by engineers or military strategists. It will be guided, or distorted, by the choices of lawmakers, ethicists, technologists, and the citizens who hold them accountable. The sky above us is filling with intelligence—watching, recording, calculating. Whether that intelligence becomes a guardian or a ghost depends on the frameworks we build today.

The choice lies not in the technology, but in how we govern it, who controls it, and for what purpose it is deployed. The sky ahead is wide open—but whether it becomes a theater of progress or peril will depend on the wisdom of the hands—and minds—that guide it.

Chapter 7: The Drone Horizon – Emerging Trends, Expanding Frontiers, and the Road Ahead

As we look beyond 2025, the drone landscape is rapidly transforming—from a tool of tactical warfare and niche industrial use into a ubiquitous layer of daily life. Unmanned aerial systems are no longer just flying cameras or delivery bots—they are becoming a critical part of the digital-physical infrastructure that connects people, places, and data in real time.

The horizon ahead is shaped by unprecedented technological innovation, policy adaptation, and societal integration. From agricultural sustainability to medical logistics, from aerial taxis to autonomous swarms, the next decade promises to bring drones closer to the ground—and deeper into the rhythms of modern civilization.

Drones are being reimagined not just as flying sensors, but as integrated problem-solvers across key sectors:

Agriculture: Drones equipped with multispectral imaging and AI diagnostics help farmers monitor crop health, detect pest outbreaks, and apply precise treatments. In regions like sub-Saharan Africa and Southeast Asia, this technology is transforming smallholder farming into a data-driven enterprise, increasing yields and reducing resource waste.

Healthcare: In partnership with organizations like Zipline and UNICEF, drones are now delivering vaccines, blood, and lab samples to rural clinics across Rwanda, Ghana, and parts of the United States. In crisis zones, UAVs are becoming the first responders, bridging critical supply chain gaps when roads are blocked or infrastructure is down.

Infrastructure Inspection: Drones are increasingly relied upon to inspect power lines, wind turbines, oil pipelines, and bridges, reducing human risk and improving inspection frequency. AI-powered platforms can now automatically detect rust, cracks, and heat anomalies from drone footage—replacing hours of manual analysis with minutes of automated reporting.

Logistics and Delivery: Amazon, UPS Flight Forward, and Wing are piloting drone deliveries for consumer goods and medications in urban and suburban settings. These initiatives aim to reduce last-mile emissions, traffic congestion, and delivery times, especially in densely populated or underserved areas.

Disaster Response: Drones have become indispensable tools for assessing damage after hurricanes, wildfires, and earthquakes. Equipped with thermal sensors and real-time mapping software, they support search and rescue operations, assist in coordination of relief efforts, and even drop emergency supplies into isolated zones.

The pace of drone innovation today is nothing short of breathtaking. What began as remote-controlled tools of surveillance has evolved into a rapidly expanding ecosystem of intelligent, autonomous machines reshaping how we interact with the world. A convergence of artificial intelligence, miniaturized sensors, next-generation batteries, and swarm behavior has propelled drones into a new era — one defined by endurance, autonomy, and purpose.

Artificial Intelligence now allows drones to think on the fly. No longer dependent on human pilots, today's UAVs can detect and avoid obstacles, identify specific targets, land with precision, and reroute themselves in real time. These capabilities, once confined to military labs, are now embedded in consumer and industrial platforms, thanks to edge-based computing and increasingly powerful onboard processors.

Swarm intelligence adds another layer of evolution. Modeled after the collective behavior of birds and insects, drone swarms can divide tasks, share data, and move as a coordinated unit — ideal for search-and-rescue missions, large-area mapping, or military operations involving ISR (intelligence, surveillance, reconnaissance) and electronic warfare. Civilian uses are equally transformative, from firefighting support to wide-field agricultural diagnostics.

Battery breakthroughs are pushing the boundaries of flight. Solid-state batteries, hydrogen fuel cells, and ultralight composites are extending flight times from minutes to hours — even days. This opens new possibilities for persistent oceanic patrols, atmospheric monitoring, and beyond-line-of-sight communications in remote areas.

Sensor payloads are also becoming more sophisticated. Today's drones carry everything from LiDAR and hyperspectral cameras to radar altimeters and AI-accelerated visual recognition systems. A single pass over a forest or mining site can generate a high-fidelity 3D model, complete with data integration into digital twin platforms. Drones are no longer just collecting data — they are contextualizing it.

As cities grow ever more congested and their roadways strain under the pressure of urbanization, the air above is being reimagined as the next great frontier in transportation. What was once a realm reserved for helicopters and the occasional news drone is now poised to become a structured aerial ecosystem—one that promises to transform the way people and goods move through metropolitan life. This vision, known as Urban Air Mobility, is no longer a flight of fancy but a rapidly developing architecture of technology, infrastructure, and regulation. In this future, electric vertical takeoff and landing aircraft—eVTOLs—shuttle passengers across cities in minutes instead of hours. Autonomous drones navigate through structured corridors to deliver medicine, packages, and even defibrillators to those in need. Aerial platforms conduct real-time surveillance, manage traffic flows, and assist emergency responders long before traditional units arrive.

Cities are beginning to adapt to this airborne transformation. Rooftops, parking garages, and dedicated towers are being repurposed as vertiports—the future's takeoff and landing stations. Coordinating this new layer of mobility are Unmanned Traffic Management systems, powered by artificial intelligence and high-speed wireless networks, which chart precise flight paths, deconflict airspace, and ensure safety amid the buzz of increasing aerial activity. Urban planners are already drafting the airspace equivalent of highways and intersections, integrating drones into the digital fabric of the smart city.

This vision is not theoretical. It is unfolding now. Companies like Joby Aviation, Lilium, and Archer are flight-testing electric air taxis capable of traveling dozens or even hundreds of miles. Volocopter has successfully demonstrated autonomous-capable vehicles ideal for short hops within urban centers. Governments and regulatory bodies are scrambling to catch up. NASA and the Federal Aviation Administration are laying the groundwork for future urban airspace standards, while nations like South Korea, Germany, China, and the United Arab Emirates are testing drone corridors in live environments. Cities from Los Angeles to Singapore are already embedding Urban Air Mobility into their development blueprints, preparing for an age of aerial prescriptions, rooftop commutes, and airborne law enforcement.

The allure of Urban Air Mobility is clear. It offers a chance to bypass gridlock, speed up emergency response, extend services to underserved communities, and reduce carbon emissions through electric propulsion. But every transformative technology brings with it a suite of new risks and dilemmas. Noise pollution from fleets of whirring rotors could quickly fray public tolerance. Navigating crowded skies above city streets demands an unprecedented level of airspace safety and coordination. Autonomous aircraft and interconnected traffic systems present tempting targets for cyberattacks. And there remains the question of access—who will benefit from this leap into the sky, and who will be left grounded?

By the early 2030s, many experts predict that a latticework of drone corridors will stretch above major urban centers, coordinating fleets of piloted and autonomous aircraft. Dispatchers may launch drones with the same routine efficiency as first responders roll out fire engines. What once belonged to science fiction—sky taxis, aerial police patrols, autonomous cargo fleets—will become woven into the ordinary functioning of city life. But Urban Air Mobility is more than a logistical or technological shift; it challenges how cities think about space, equity, and governance. As aerial infrastructure matures, it will reshape building codes, alter emergency planning, and test civil liberties in ways society is only beginning to grasp.

To safely integrate this sky-bound future, regulation must evolve just as swiftly as the aircraft themselves. Agencies like the FAA, EASA, and ICAO are now racing to define a coherent set of rules that prioritize public safety while enabling innovation. One cornerstone of this effort is Remote Identification—a mandate requiring drones to broadcast their location, altitude, and operator identity in real time, allowing for tracking and accountability in crowded skies. Equally critical is the expansion of Beyond Visual Line of Sight (BVLOS) operations, which would permit drones to fly extended routes without constant human supervision, enabled by advanced detect-and-avoid systems and AI-guided navigation. These advances, in turn, will feed into the growing need for comprehensive Unmanned Traffic Management networks, capable of coordinating thousands of flights simultaneously across urban airspace.

International cooperation is becoming increasingly essential. As drones cross borders and skies fill with international fleets, standardized protocols for licensing, airworthiness, and data sharing are necessary to ensure global interoperability. Organizations like ICAO and JARUS are working to harmonize regulations across regions, but the pace of progress varies widely by country and political will.

Beyond regulatory logistics lie deeper ethical and societal questions. As drones become embedded into daily life—watching, recording, and reacting in real time—privacy rights are increasingly at risk. Persistent aerial surveillance by police, corporations, or individuals poses a direct challenge to expectations of anonymity and freedom in public and semi-private spaces. New laws must be crafted to ensure transparency, limit data retention, and protect citizens from unaccountable monitoring.

Automation will also bring economic disruption. Jobs in delivery, infrastructure inspection, and mapping—once the domain of human labor—are increasingly at risk from drones. While UAVs can improve efficiency and reduce danger in certain fields, policymakers must plan for retraining programs and workforce transitions to cushion the social impact.

Even the environmental benefits of drones must be measured against the full lifecycle cost of their manufacture, battery production, and potential disruption to wildlife. Sustainable design, recyclability, and intelligent flight planning will become priorities as fleets scale up.

In the end, the promise of the aerial city cannot rest solely on engineering. It must be matched by foresight, ethics, and equitable design. The future of Urban Air Mobility is not just a new way to move through space—it is a redefinition of that space itself. If done right, the sky will not just be the limit, but a networked extension of the city—responsive, inclusive, and secure. If done poorly, it risks becoming yet another domain where inequality, surveillance, and congestion take flight. The challenge now is to guide this ascent with wisdom as well as ambition, ensuring that what rises above us reflects not only our technological capability but our collective values.

Looking to the horizon, drones are poised to define what many are calling the "third dimension of civilization." As cities grapple with gridlocked roads and terrestrial infrastructure reaches its limits, the sky above is becoming a critical new layer of human activity. No longer just a realm for birds and aircraft, it is transforming into an aerial corridor for commerce, communication, transportation, and even warfare. This evolving stratum of society—intelligent, autonomous, and increasingly indispensable—will carry not just goods and people, but also data, services, and decisions.

Urban air mobility stands at the vanguard of this transformation. Companies like Joby Aviation, Volocopter, and EHang are actively shaping a future in which electric aerial taxis and autonomous cargo drones shorten commutes, revolutionize logistics, and redefine how cities are structured. Vertiports will become as commonplace as subway stops, and aerial routes will stretch above skylines as seamlessly as bus lines thread through neighborhoods today. What was once fantasy is becoming integrated infrastructure, tested and mapped in real-world environments.

At the same time, drones are rapidly becoming instruments of next-generation warfare. Nations are investing heavily in AI-powered uncrewed fleets—capable of autonomous target recognition, adaptive flight formations, and lethal action without direct human input. These developments have ignited urgent international debates about ethics, accountability, and the risks of an AI arms race spiraling out of human control. In this high-stakes frontier, strategic advantage could hinge not on firepower, but on algorithms.

Equally transformative is the rise of bio-inspired and nano-scale drones. Engineers are now drawing on nature's designs to create surveillance bots that resemble dragonflies, beetles, or hummingbirds. These insect-sized platforms can crawl walls, enter buildings undetected, or hover in dense environments that confound traditional aircraft. While their applications in search-and-rescue and reconnaissance are promising, they also challenge societal norms around privacy and consent—raising questions that few legal systems are prepared to answer.

In an era defined by disruption—be it from climate disasters, pandemics, or geopolitical shocks—drones offer resilience through autonomy. Able to operate without putting human lives at risk, they have become essential tools in supply delivery, medical logistics, disaster mapping, and emergency communications. Yet with this newfound independence comes the imperative for oversight. Autonomy must not outpace responsibility; the systems that fly must remain tethered to ethical frameworks that protect civil liberties and human life.

The future of drones is not a far-off abstraction. It is already unfolding above our cities, across our battlefields, and along our supply chains. The airspace once defined by altitude alone is now layered with meaning and purpose, shaped by code, connectivity, and consequence. We are building not just new machines, but new norms. And the choices we make—about what we permit to fly, who has access, how decisions are made, and what safeguards we implement—will resonate across generations.

The sky is no longer empty. It is becoming a platform, a marketplace, a battlefield, and a commons. The challenge before us is not simply to master flight, but to ensure that what ascends into our skies reflects our highest values. Innovation must be matched by governance, and technological ascendancy by moral clarity. The third dimension is here—and it is ours to shape.

Chapter 8: Urban Skies – Drones and the Smart City Revolution

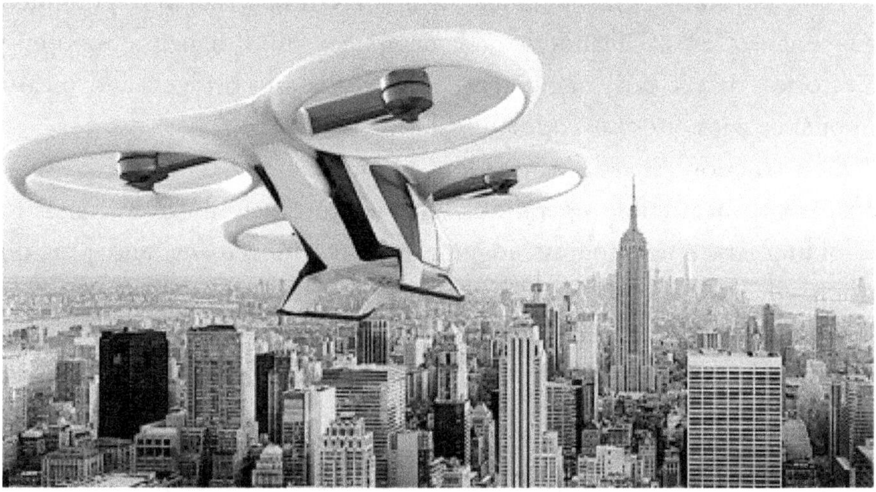

As the global urban population marches toward an estimated 70% by 2050, cities are confronting immense challenges in sustaining infrastructure, delivering essential services, and maintaining environmental balance. In response, the world's urban centers are undergoing a profound transformation—an evolution into "smart cities" that leverage cutting-edge technologies to optimize their operations and enhance the lives of their inhabitants. At the forefront of this revolution is a technology that, not long ago, seemed more like science fiction than city planning: the drone.

Unmanned aerial vehicles have quickly moved from novelty to necessity, becoming a critical component in how cities function, evolve, and respond. Their agility and versatility allow them to serve as the eyes, ears, and even hands of city planners and emergency managers. Drones offer a dynamic perspective—hovering above congested streets, weaving between buildings, and capturing granular data that traditional methods simply cannot. What they provide is not just an overhead view, but an evolving narrative of the urban environment in motion.

One of the most transformative applications of drones is in infrastructure monitoring and urban development. Outfitted with high-resolution imaging systems, LiDAR arrays, and multispectral sensors, these machines can scan entire districts in hours, mapping surfaces with precision and constructing detailed 3D models from aerial photogrammetry. They enable urban planners and engineers to visualize development trajectories, detect early signs of wear in roads and bridges, and measure compliance with building codes in real time.

Construction sites are no longer opaque zones of activity but transparent, measurable operations. Drones routinely fly over cranes and scaffolding, tracking progress, logging material movement, and providing automated documentation for project managers. In the aftermath of earthquakes or storms, drones take to the skies before rescue teams even arrive, delivering high-fidelity maps of the damage zone, identifying access routes, and locating survivors.

In cities built on fault lines, like Tokyo or San Francisco, drones have become integrated into resilience protocols. When a quake hits, it's not just first responders who deploy—it's also fleets of UAVs programmed to fly predetermined routes, capture structural data, and beam it back to command centers. This immediate situational awareness accelerates response time, optimizes resource deployment, and saves lives.

As cities grow more complex, the role of drones is shifting from support to synergy. They no longer serve simply as tools but as essential components of an urban ecosystem that senses, reacts, and adapts. In the age of smart cities, the drone is not just a flying camera—it is a vital node in the nervous system of a metropolis increasingly defined by real-time intelligence and autonomous coordination. The urban drone era has arrived, and with it, a new way to see, shape, and sustain the cities of tomorrow.

The future of urban aviation is taking shape above our cities, where the hum of engines is giving way to the quiet whir of electric rotors. At the heart of this transformation are electric vertical takeoff and landing aircraft—eVTOLs—designed not merely to replace helicopters, but to usher in a radically different paradigm of aerial transportation. Unlike their gas-powered predecessors, eVTOLs are engineered for efficiency, sustainability, and seamless integration into the fabric of daily urban life. They represent the next evolutionary step in how cities move.

Developers like Joby Aviation, backed by NASA partnerships, are flight-testing sleek, low-noise aircraft capable of shuttling passengers from rooftop to rooftop in minutes. Hyundai's Supernal project envisions scalable, autonomous air mobility solutions that plug directly into smart city infrastructure. Eve Air Mobility, emerging from aviation giant Embraer, is pursuing a global network of air taxis tailored to the commuting needs of megacities. Together, these companies are working not just to fly—but to fundamentally reshape the urban commute.

The promise is seductive: slashing travel times, bypassing gridlock, and connecting distant corners of sprawling cities with an elevated network of clean, quiet aircraft. But this vision will only become reality through an equally ambitious rethinking of infrastructure and policy. Vertiports—purpose-built landing and takeoff zones—must be integrated into rooftops, garages, and transit hubs. Charging stations, outfitted with fast-swap batteries or renewable energy tie-ins, must ensure continuous uptime. And most critically, advanced air traffic systems must emerge to coordinate this aerial ballet. These systems will rely on AI-driven Urban Air Mobility Traffic Management platforms, blending satellite tracking, 5G communications, and edge computing to safely orchestrate thousands of simultaneous flights.

Already, cities like Dubai, Los Angeles, and Singapore are preparing for this future by piloting eVTOL corridors, designing sky lanes, and testing regulatory frameworks. Their airspaces are becoming laboratories for what may soon be the global norm: a layered transportation network where the vertical axis is no longer reserved for emergencies or elite travel, but democratized as a commuter solution.

Alongside passenger transport, drones are quietly revolutionizing another vital segment of the urban economy—the last mile. That final leg of delivery, often the most costly and inefficient, is being streamlined by unmanned aerial systems that skip traffic lights and roadblocks altogether. In smart cities, these drones are already carrying blood transfusions between hospitals, delivering vaccines to remote clinics, and dropping parcels in hard-to-reach neighborhoods during peak congestion hours.

Companies like Wing, Zipline, DroneUp, and Manna are not just experimenting—they're scaling. Their urban drone networks are proving that logistics can be faster, greener, and more resilient. When floods, wildfires, or pandemics bring conventional delivery systems to a halt, drones offer a lifeline—autonomous, trackable, and immune to street-level disruption.

But success in this domain hinges on more than just flight mechanics. Noise abatement must be addressed so that fleets of buzzing UAVs don't alienate the very communities they're designed to serve. Precision navigation and secure payload systems are essential to ensure accurate deliveries and prevent loss or tampering. And above all, public trust must be earned—through transparent policies that protect privacy, limit surveillance overreach, and clearly communicate how and where drones operate.

This convergence of eVTOL air taxis and delivery drones is not just reshaping transportation—it's redefining the urban experience itself. The sky is no longer empty space above us. It is becoming a navigable, programmable layer of infrastructure, capable of lifting not only goods and people, but the very potential of what cities can be.

Uncrewed aerial vehicles (UAVs), or uncrewed aerial systems (UASs), are revolutionizing environmental monitoring by enabling a new generation of ultra-high-resolution surveys in both space and time. Drones can bridge the longstanding gap between satellite imagery, which often lacks granularity, and ground-based fieldwork, which is time-consuming and spatially limited.

By flying low and capturing data in real time, UAVs provide ecologists, environmental agencies, and urban planners with dynamic tools to monitor ecosystems, track changes, and respond to threats with unprecedented speed and accuracy.

Unmanned Aerial Vehicles (UAVs) are transforming how we understand, monitor, and manage the planet's most vital landscapes. From mountain ranges to farmlands, these aerial tools have become indispensable in environmental science and sustainable agriculture, offering precision, reach, and insight once impossible through ground-based observation.

Equipped with advanced imaging systems such as photogrammetry and LiDAR, drones can map terrain in fine detail, generating high-resolution orthomosaics, digital surface models (DSMs), and three-dimensional reconstructions of the earth's surface. These models serve as dynamic tools for tracking changes over time — revealing the slow creep of soil erosion, shifts in vegetation density, or the reconfiguration of landforms due to natural forces or human activity. In fields from geology to watershed management, UAVs have become the cartographer's brush in the sky.

With multispectral and hyperspectral sensors onboard, drones are enabling scientists to monitor ecosystems with a level of clarity and consistency never before possible. These tools can identify biodiversity hotspots, trace the paths of migratory species, and locate critical habitats for endangered wildlife. From above, UAVs reveal patterns that can go undetected at ground level — changes in forest canopy, the encroachment of invasive species, or early signs of habitat degradation. Their insights are already guiding conservation efforts, helping resource managers prioritize interventions, protect sensitive zones, and plan reforestation or habitat restoration with pinpoint accuracy.

In the agricultural realm, drones are helping farmers turn data into yield. By flying over fields and capturing spectral data, UAVs can assess plant health using vegetation indices such as NDVI (Normalized Difference Vegetation Index). These indices expose invisible stress — drought, nutrient deficiencies, pest infestation — long before symptoms appear to the naked eye. With this intelligence, farmers can make targeted decisions: applying fertilizer where it's needed, conserving water in over-irrigated zones, or deploying treatments only where pests are detected. The result is a new form of agriculture — one that is more efficient, environmentally responsible, and technologically empowered.

This **data-driven approach**, known as precision agriculture, does more than improve productivity — it increases yield while reducing water consumption, chemical runoff, and soil depletion. It marks a shift from broad-stroke farming to highly targeted land stewardship, where efficiency and environmental health go hand in hand.

Beyond the fields, drones are revolutionizing how we observe and model water systems. Using advanced image velocimetry, UAVs capture surface flow velocities across rivers and streams, allowing scientists to construct accurate two-dimensional flow fields. These models are essential not only for predicting floods and analyzing sediment transport but also for assessing the health of aquatic habitats. With the ability to monitor seasonal changes or capture real-time surge events, drones are fast becoming indispensable tools in hydrology and watershed science.

In environments that are too dangerous or difficult for manual inspection, drones provide a safer alternative. High-resolution imaging and thermal sensors allow UAVs to assess the structural integrity of critical infrastructure — including bridges, dams, power stations, and railway corridors — without risking human life. By flying close to inaccessible surfaces, drones can identify stress fractures, material fatigue, or early signs of corrosion, enabling proactive maintenance and avoiding catastrophic failure.

In mining regions, drones equipped with hyperspectral cameras are changing the landscape of environmental monitoring. These sensors can detect acid mine drainage, track chemical residues, and identify proxy minerals such as goethite and jarosite — key indicators of toxic pH levels. By mapping these signatures from above, drones assist in both the assessment and remediation of contaminated sites. In post-mining reclamation zones, they offer a non-invasive way to monitor recovery, target intervention, and ensure compliance with environmental regulations.

As cities become denser and more digitally interconnected, drones are emerging as essential tools for managing urban ecosystems, safeguarding public health, and enhancing emergency response. These airborne systems offer a dynamic perspective on complex metropolitan challenges — from environmental monitoring to disaster relief.

Urban air pollution remains a silent threat. Drones equipped with advanced chemical sensors are now able to detect airborne pollutants with precision — including particulate matter (PM2.5 and PM10), carbon dioxide (CO_2), methane (CH_4), and volatile industrial emissions. By sampling the atmosphere across neighborhoods and industrial zones, these UAVs generate real-time air quality data that feeds directly into public health dashboards. City planners and environmental agencies can use this information to enforce emissions regulations, issue health advisories, and adapt infrastructure in the face of climate change.

From above, drones reveal what ground inspectors often miss. Aerial surveys detect illegal dumping sites, monitor landfill expansion, and identify inefficiencies in waste handling and recycling operations. With regular flyovers, municipalities can track the environmental footprint of their waste systems, prioritize clean-up efforts, and ensure that recycling programs remain accountable and transparent.

In cities struggling with rising temperatures, drones equipped with thermal imaging cameras are mapping heat islands in stunning detail. These hotspots — areas where asphalt, concrete, and steel trap and radiate heat — can be visualized at scale, helping officials pinpoint neighborhoods most vulnerable to extreme heat. Armed with this data, cities can implement targeted mitigation strategies such as tree planting, green roofing, reflective surfaces, and the installation of cooling infrastructure.

In moments of crisis, drones become lifelines. During wildfires, floods, earthquakes, or major infrastructure failures, UAVs provide aerial views that ground teams cannot obtain in real time. They help assess damage, locate survivors, monitor evacuation routes, and coordinate rescue missions — all while keeping first responders at a safe distance. In emergencies where every second counts, drones deliver eyes in the sky that can make the difference between chaos and coordination.

For public safety agencies, drones are becoming indispensable tools for maintaining order and situational awareness. Deployed above parades, protests, sports events, or urban bottlenecks, UAVs provide live feeds to command centers, allowing officers to monitor crowd density, detect hazards, and direct foot or vehicle traffic more effectively. In evacuation scenarios, they help trace optimal escape routes and minimize confusion. With faster deployment than helicopters and a much smaller footprint, drones enhance public safety while reducing personnel risk.

With the convergence of cutting-edge sensors, AI-powered analytics, and high-speed 5G communications, drones are emerging as vital components of next-generation environmental monitoring systems. In both remote wilderness and high-density urban zones, UAVs now serve as airborne sentinels—tracking pollution, mapping floodplains, detecting wildfires, and measuring air quality in real time. Their mobility and flexibility allow for faster responses to ecological threats and more informed decision-making in sustainability planning. These flying platforms can now deploy thermal cameras to scan for heat anomalies in forests, sniff for methane leaks near pipelines, and even assist in biodiversity monitoring by identifying plant health or animal movement across large, inaccessible terrains.

In the urban sphere, drones are increasingly woven into the fabric of climate adaptation and smart city architecture. They support urban heat mapping, monitor water runoff during storms, and even survey green rooftops and solar installations. The vision is clear: a responsive, data-driven infrastructure where UAVs act as eyes in the sky for city planners, emergency responders, and environmental scientists alike.

Yet, as this vision expands, it must be navigated with care. The very capabilities that make drones so powerful also introduce new ethical, legal, and social complexities. The ability to collect visual, thermal, and in some cases biometric data, demands robust frameworks to define what drones are permitted to observe, how long data can be retained, and under what circumstances it can be shared. Without these protections, the public risks becoming subjects of invisible surveillance, eroding trust in both government and technology.

Security is another critical concern. As drones become more capable and autonomous, they also become potential vectors for disruption. A hacked UAV, whether repurposed for surveillance or equipped with malicious payloads, could threaten public safety at events, damage critical infrastructure, or interfere with communications systems. The defense against such threats requires not just technical safeguards like encrypted links and real-time geofencing, but also responsive legal structures and enforcement tools.

Regulatory clarity is essential. National aviation agencies are working to develop overarching frameworks, but cities must also define local rules that reflect their unique geography, demographics, and risk profiles. This includes establishing no-fly zones, integrating Remote Identification (RID) systems that broadcast drone identity and location in real time, and setting altitude and traffic limits to prevent collisions and overcrowding.

At the center of all this is the issue of public trust. The social license to operate UAVs in public spaces cannot be assumed—it must be earned. Cities and drone operators must prioritize transparency, inviting community input on where, when, and why drones are deployed. Public forums, data access portals, and civilian oversight mechanisms can help ensure accountability. Without these safeguards, even well-intentioned drone programs could provoke backlash, derailing progress just as UAVs reach their full potential.

Ultimately, the future of drones in urban ecosystems will hinge not only on what the machines can do—but on how wisely they are governed. A sky filled with intelligent machines offers extraordinary promise, but only if that promise is shaped by ethical foresight, democratic input, and responsible design.

To truly harness the power of drones, cities must stop viewing them as gadgets and start recognizing them as **critical infrastructure** — no less foundational than fiber optics, water mains, or public transportation. As unmanned aerial systems become more embedded in civic operations, they are evolving into the connective tissue of the smart city: fusing mobility, data, safety, and sustainability into a dynamic urban network.

But realizing this potential will require more than technology. It demands foresight, coordination, and governance. **Interagency collaboration** will be essential — not just between aviation regulators, but across departments of transportation, public health, emergency management, and urban planning. Drones touch every layer of city life, and no single agency can manage their deployment in isolation.

Cities must also embrace **smart zoning for aerial corridors** — designated layers of airspace that separate piloted air taxis, autonomous delivery drones, and recreational UAVs. Just as roads are divided by speed and function, the skies above urban centers will need structured pathways, supported by Unmanned Traffic Management (UTM) systems and live airspace analytics.

Public-private partnerships will be the engine of innovation. Municipal governments must collaborate with drone developers, AI firms, and logistics providers to pilot new applications, scale up what works, and ensure that commercial interests align with public needs. These partnerships will determine whether aerial systems serve all residents — or only the wealthiest and best connected.

Finally, none of this progress will matter without a foundation of **ethical design and inclusive governance**. Drones must be deployed with respect for civil liberties, community consent, and environmental sustainability. Access must be equitable. Oversight must be real. And citizens must be more than passengers — they must be participants in shaping the rules of this new aerial domain.

In short, drones are helping to write the next chapter of urban civilization. From curb to cloud, from crisis response to everyday logistics, they are no longer just tools — they are infrastructure. And how we choose to integrate them will define what kind of cities we live in tomorrow.

Chapter 9: Global Players and the Drone Arms Race

As drone technology continues to evolve, its reach has expanded far beyond the engineering labs of Silicon Valley and the black-budget hangars of Western defense contractors. Unmanned aerial vehicles have matured into instruments of global significance—reshaping economic strategies, redefining military doctrine, and raising new questions about sovereignty, airspace, and the ethics of automation. The twenty-first century's geopolitics are increasingly unfolding not only on land or sea, but in the skies—occupied not just by manned aircraft, but by fleets of intelligent, autonomous machines.

Nowhere is this transformation more visible than in the expanding international landscape of drone integration. The United States, long a pioneer in UAV development, remains at the forefront of both commercial innovation and military experimentation. Domestically, the Federal Aviation Administration has led a steady progression from cautious oversight to adaptive regulation. With mandatory Remote ID protocols now in place, drones operating in U.S. airspace must broadcast identifying and positional data—laying the groundwork for more complex operations to follow. By 2025, new FAA guidelines are expected to further liberalize Beyond Visual Line of Sight (BVLOS) operations, a move that could revolutionize sectors such as industrial inspection, agriculture, and delivery logistics. Meanwhile, in the defense arena, the Pentagon continues to invest in next-generation autonomous systems. Programs like the XQ-58A Valkyrie, a stealthy, high-speed combat drone, the LongShot air-launched weapon system, and DARPA's Gremlins initiative for reusable UAV swarms are shaping a future where unmanned platforms fly in formation with crewed aircraft, acting as decoys, scouts, and strike assets in contested environments.

North of the border, Canada is advancing its own drone agenda at a measured pace. The introduction of the Remote Pilot Operator Certificate (RPOC), combined with a new suite of pilot certification exams slated for implementation in 2025, reflects an effort to formalize UAV competency nationwide. By the end of that year, Canada plans to authorize routine

BVLOS flights, unlocking critical drone applications across its expansive and often inaccessible terrain. From monitoring remote oil pipelines to delivering medicine to Indigenous communities in the Arctic, drones are fast becoming essential tools in bridging the country's geographic and service divides.

Across the Atlantic, the European Union has emerged as a regulatory pioneer. The European Union Aviation Safety Agency (EASA) has developed a nuanced framework that categorizes drones by their level of operational risk—Open, Specific, or Certified—allowing regulators and operators alike to tailor requirements to the nature of each mission. This risk-based model enables flexible yet safe drone deployment, balancing innovation with accountability. At the same time, the EU's investment in U-space—the continent's vision for a coordinated Unmanned Traffic Management (UTM) ecosystem—signals its commitment to managing high-density UAV traffic at scale. Through the SESAR initiative, Europe is actively funding pilot corridors and cross-border drone applications, from intercity cargo flights to real-time wildfire surveillance.

Following its departure from the EU, the United Kingdom has charted a parallel course. The UK Civil Aviation Authority (CAA) has taken on full responsibility for UAV regulation, with upcoming 2026 policies aimed at streamlining market surveillance of drones under 25 kilograms. However, British efforts to field advanced military UAVs have encountered domestic friction. Strict privacy protections and airspace constraints have complicated live-fire exercises and multinational training missions, prompting concern that the UK may lag behind its allies in real-world combat readiness for unmanned systems.

Together, these developments reflect a rapidly diversifying drone ecosystem—one in which national strategies diverge not only in pace and policy, but in purpose. While Western democracies grapple with balancing innovation and oversight, the broader international scene reveals a more complex and contested future. In the chapters ahead, the emergence of new drone powers, the influence of non-state actors, and the simmering technological arms race between global superpowers will only deepen the strategic implications of an unmanned world.

China has firmly established itself as the dominant force in the global drone industry, leading both consumer and military markets with an expansive portfolio of unmanned systems. Flagship companies such as DJI, Ziyan, and the China Aerospace Science and Technology Corporation (CASC) have become synonymous with high-volume, cost-effective, and increasingly capable UAV platforms. Beijing's Military-Civil Fusion strategy has supercharged this trajectory, channeling academic research, private-sector innovation, and military funding into unified drone development pipelines. The result is a rapidly maturing ecosystem focused not only on individual drone performance, but on advanced capabilities like swarm coordination, AI-driven target acquisition, and precision-guided loitering munitions.

Domestically, China has moved aggressively to regulate and monitor its aerial environment. UAV registration, operator licensing, and dynamic airspace zoning are all in place, forming a highly controlled drone governance model. Internationally, the country's combat drone exports—particularly the CH-4 Rainbow and Wing Loong families—have proliferated widely. These platforms are now key components of the arsenals of countries such as Pakistan, Egypt, Saudi Arabia, and the United Arab Emirates. By offering armed UAVs without the political strings typically attached to Western exports, China has expanded its influence across key geostrategic corridors in the Middle East, South Asia, and North Africa.

India, meanwhile, has embarked on a broad-spectrum drone adoption strategy, deploying UAVs across defense, agriculture, logistics, and public administration. At the heart of this effort lies the Digital Sky Platform, an ambitious digital infrastructure overseen by the Directorate General of Civil Aviation. This online portal enables real-time drone registration, geo-fencing compliance, and automated flight approval. On the defense front, India's state-run institutions and private partners are manufacturing homegrown tactical drones such as the Rustom series and the stealthy Ghatak UCAV prototype. Concurrently, Indian startups are advancing drones for precision farming, medical deliveries to remote regions, and infrastructure inspection across the country's vast and diverse terrain.

In the Asia-Pacific, Australia has taken a structured and safety-first approach to drone integration. Its Civil Aviation Safety Authority mandates registration for any drone over 250 grams and requires operational approval for commercial use. The nation's unique geography has made UAVs invaluable in sectors such as mineral exploration, bushfire response, and ecological monitoring. Australia has also made notable strides in defense UAVs through its joint program with Boeing on the Ghost Bat, a semi-autonomous "loyal wingman" drone designed to fly alongside crewed fighter jets—representing a bold step into cooperative combat systems.

In Africa, drones have proven transformational in ways distinct from the high-tech arms races of superpowers. Countries such as Rwanda, Ghana, and Malawi are leveraging UAVs to overcome legacy infrastructure challenges. Rather than laying costly roads or building supply depots, health ministries now partner with firms like Zipline to deliver blood, vaccines, and lab samples to rural clinics in minutes instead of hours. These humanitarian UAV corridors—once dismissed as aspirational—have become real-world models of what aerial logistics can achieve in under-resourced settings. Still, widespread adoption across the continent faces barriers, including the high cost of equipment, underdeveloped legal regimes, and limited air traffic coordination infrastructure.

In Latin America, drone use reflects both the region's agricultural strength and its uneven regulatory development. Countries like Brazil, Uruguay, and Mexico have embraced UAVs in precision farming, deforestation tracking, and police operations. Yet progress is patchy. While Brazil has pushed forward with industrial and agritech drone deployment, others lag due to outdated aviation statutes or bureaucratic inertia. In the Caribbean, drone usage remains largely recreational or commercial, with applications focused on tourism media, aerial mapping, and disaster assessment—valuable, but not yet systemically integrated.

What emerges from this global survey is a mosaic of drone integration—highly advanced in some regions, emergent and localized in others. As UAVs continue to evolve and proliferate, the gaps between nations in terms of capability, policy, and ethical oversight may well define future geopolitical asymmetries as clearly as differences in GDP or military size. The drone age is not only arriving—it is arriving unevenly.

In the Middle East, drone proliferation has accelerated dramatically, with countries like the United Arab Emirates, Saudi Arabia, and Israel aggressively expanding their UAV capabilities for both civilian and military purposes. These nations view unmanned systems not merely as tools of tactical advantage, but as strategic assets critical to sovereignty, security, and technological leadership.

Israel, long recognized as a pioneer in UAV innovation, remains at the forefront with an array of battle-tested platforms such as the Heron, Harop, and Hermes series. These drones have been extensively used in surveillance, precision strikes, and electronic warfare. Israel's doctrine emphasizes seamless integration of UAVs into ground and air operations, enabling real-time intelligence and surgical targeting that minimize traditional troop deployments. Its UAVs are not just deployed domestically—they are exported globally, forming the backbone of many smaller nations' drone fleets.

The United Arab Emirates has taken a dual-pronged approach: investing in autonomous drone logistics for urban deliveries and emergency response, while also building military capabilities through partnerships and acquisitions. Dubai has become a testbed for smart city drone integration, hosting trials for air taxis and drone delivery corridors, while Abu Dhabi focuses on defense procurement and indigenous development through its EDGE Group.

Saudi Arabia, while historically dependent on foreign platforms, is rapidly building its domestic drone industry through joint ventures with nations like China and South Africa. Riyadh's Vision 2030 strategy includes a focus on defense self-sufficiency, and UAVs play a central role in this ambition. The kingdom is pursuing platforms for border surveillance, internal security, and drone countermeasures—particularly in response to recent attacks on oil infrastructure.

Meanwhile, Iran has embraced a different model—prioritizing low-cost, long-range drones designed for asymmetric warfare and regional influence. The Iranian arsenal includes platforms like the Shahed-129, Mohajer-6, and a growing class of kamikaze-style loitering munitions. These drones are not only deployed directly but are also exported to proxy forces such as Hezbollah in Lebanon and the Houthis in Yemen. Their missions range from battlefield reconnaissance and artillery spotting to precision strikes on strategic infrastructure. The 2019 attacks on Saudi Aramco facilities by drones attributed to Houthi or Iranian origin marked a watershed moment, demonstrating how inexpensive UAVs could bypass traditional air defenses and cause economic disruption on a massive scale.

Iran's strategy is not just military—it is geopolitical. By proliferating UAV technology to allied non-state actors, Tehran has created a regional web of drone-enabled proxy warfare. This approach allows Iran to challenge its adversaries across multiple fronts without incurring the direct costs or risks of conventional warfare. It represents a potent example of how drone technology can rebalance power dynamics in contested regions.

At the global level, the drone race is increasingly a proxy battlefield in the broader U.S.-China technology rivalry. In early 2025, Beijing imposed retaliatory export controls targeting 28 American defense and aerospace firms, a direct response to tightening U.S. restrictions on Chinese-made UAVs. These tit-for-tat moves signal an intensifying techno-nationalist conflict in which drones are not just hardware, but vessels of data, influence, and strategic autonomy.

The United States has responded with proposed legislation to ban all federally funded purchases of Chinese drones, citing the 2017 Chinese intelligence law that compels firms like DJI to share data with state authorities. Given DJI's overwhelming dominance in the civilian drone market—controlling over 70% globally—this move has triggered a scramble across U.S. agencies and industries to identify domestic alternatives. Simultaneously, new investments are flowing into U.S. drone startups and manufacturing hubs aimed at reshoring production and securing supply chains against foreign influence.

This global divergence is creating a fragmented drone ecosystem. Nations once united by open supply chains and shared technical standards are now drawing hard lines in the sand—favoring homegrown platforms, trusted trade partners, and strict data sovereignty. The result is a world in which drone innovation continues to accelerate, but under increasingly nationalist terms. As UAVs become essential infrastructure for logistics, defense, surveillance, and urban mobility, the stakes of who builds, controls, and governs these systems are rising fast. The skies may be open, but the geopolitics of drones are anything but.

This emerging era of drone proliferation reflects a deepening distrust in global technology supply chains. Nations that once embraced a globalized model of innovation are now retreating into sovereign development, wary that foreign-built drones might serve as Trojan horses—quietly collecting data or vulnerable to remote interference. As tensions between global powers escalate, drone infrastructure is becoming both a symbol and instrument of national autonomy. The result is a fractured ecosystem where platforms are shaped not just by engineering constraints, but by politics, ideology, and security imperatives.

At the same time, the once-clear boundary between state and non-state military power is eroding. Armed groups, insurgents, and militias now field UAV capabilities once reserved for nation-states. Loitering munitions assembled from off-the-shelf components, modified commercial quadcopters used for grenade drops, and low-cost surveillance drones deployed for battlefield intelligence are all part of this new asymmetric airpower. These actors have become UAV-enabled micro-militaries, capable of projecting force, intimidating adversaries, and altering the tactical balance—often without facing the diplomatic or military repercussions that nation-states must consider.

In the years ahead, the question is no longer whether drones will dominate the sky, but how that dominance will be structured—under what legal, ethical, and strategic rules, and in whose hands the control of this airspace will rest. The future will be shaped not merely by technical prowess, but by the collective choices of governments, industries, and civil societies as they navigate the complex terrain of autonomy, accountability, and power in the age of unmanned flight.

Chapter 10: The Future of Drones – Sky Is No Limit

As the world accelerates into a hyper-connected, data-driven future, drones are no longer confined to the margins of innovation. They are rising swiftly to become central actors in an emerging aerial ecosystem that touches every aspect of life—from mobility and commerce to emergency response and planetary stewardship. What was once considered futuristic or speculative is now unfolding in real time, reshaping the very concept of the sky as shared, dynamic infrastructure.

Urban Air Mobility is at the forefront of this evolution. Cities long defined by the limits of roads, rails, and subways are now beginning to reimagine their skylines not only as vertical real estate, but as airborne corridors of movement. In this vision, electric vertical takeoff and landing aircraft glide quietly above the traffic-choked streets, transporting passengers and goods with unprecedented speed and efficiency. These eVTOLs—autonomous, zero-emission, and scalable—are poised to relieve pressure from aging infrastructure while offering new dimensions of accessibility.

Companies around the world are no longer just prototyping—they are preparing for deployment. Volocopter's VoloCity, with its sleek frame and whisper-quiet rotors, is designed for dense urban corridors and short intra-city routes. Boeing-backed Wisk Aero is pushing even further, developing an autonomous air taxi that eliminates the need for a pilot altogether, merging AI autonomy with commercial-grade aviation safety. Supernal, Hyundai's advanced air mobility division, and Eve Air Mobility, spun out of Embraer, are both collaborating with city governments to build the next layer of public transit: smart vertiports, aerial traffic management, and multimodal hubs that integrate seamlessly with trains, rideshares, and pedestrian networks.

This is not theoretical. Brisbane intends to make air taxis a centerpiece of the 2032 Olympic Games. In China, where national industrial strategy and innovation incentives align, plans are underway to deploy 100,000 flying vehicles within the decade—an ambitious vision that encompasses logistics, private commuting, and emergency services. These developments signal not just a shift in transportation, but a cultural transformation in how people understand and use the air above them.

Meanwhile, the advance of delivery drones continues at a breakneck pace. Purpose-built to carry everything from food and medicine to tools and mail, these UAVs are rewriting the rules of logistics. No longer bound by terrain, traffic, or fixed infrastructure, they offer an elegant solution to the perennial "last mile" challenge. For isolated clinics, mountain villages, or disaster zones cut off by floods or earthquakes, drones can mean the difference between hours and minutes, between delay and survival.

From Zipline's life-saving medical flights in Rwanda and Ghana to Wing's suburban delivery programs in Virginia and Australia, the model is already working. In cities, drones weave through air corridors to drop packages directly at homes or designated rooftops, reducing emissions and avoiding road congestion. In rural regions, they leapfrog traditional infrastructure entirely, expanding access to goods and services that once required hours of travel or were simply out of reach.

Together, Urban Air Mobility and autonomous delivery networks are converging into a seamless, intelligent skyway—one that promises to lift not just vehicles, but the very foundations of how societies move, respond, and thrive. As the airspace above becomes as orchestrated as the roads below, the sky ceases to be the limit. It becomes the next platform for civilization.

Major global corporations such as Amazon, DHL, and FedEx have become early adopters of drone-based delivery platforms, investing heavily in autonomous aerial logistics networks that promise faster and more efficient service than traditional ground transportation. While still in the developmental and regulatory phase in many regions, drone delivery systems have already transitioned from experimental novelty to functional reality in select markets.

The global response to the COVID-19 pandemic further accelerated this transformation. As conventional supply chains faced disruptions, delivery drones were deployed in dozens of countries to distribute vaccines, test kits, and critical medical supplies — often to locations inaccessible by road or in areas under lockdown. In Africa, organizations such as Zipline partnered with governments in Ghana, Uganda, and Rwanda to fly vital blood products, personal protective equipment, and vaccines across hundreds of miles of rural terrain. Similar projects launched in Australia, Canada, the United Kingdom, and the United States, validating the technology in real-world conditions and underscoring its life-saving potential.

While aerial drones are the most common form, the broader category of delivery drones may also include terrestrial (ground-based), aquatic, and even amphibious systems, depending on terrain and application. These UAVs may operate in fully autonomous, semi-autonomous, or remotely piloted modes, using advanced GPS navigation, real-time sensor data, and collision-avoidance algorithms to complete missions safely and efficiently.

Among the many commercial applications of drone technology, medical supply delivery has proven to be the most widely tested and socially impactful. Governments and humanitarian organizations have embraced this use case not only for its efficiency, but for its ethical imperative: reaching the underserved, the isolated, and the vulnerable in record time — with tools that may save lives.

As regulatory frameworks mature and urban airspace becomes more intelligently managed, delivery drones are poised to move from experimental novelty to an embedded layer of modern logistics. No longer restricted to pilot programs or disaster zones, these unmanned aerial vehicles are steadily redefining how goods are moved, emergencies are managed, and critical services are delivered—reshaping public expectations around speed, accessibility, and precision.

The groundwork for this transformation is already underway. Across global cities, investment is accelerating in vertiport infrastructure equipped with integrated charging systems and modular passenger terminals. These rooftop or roadside hubs are being designed to accommodate a mixed fleet of air taxis and delivery drones, with the same level of planning now reserved for bus terminals or train stations. In parallel, digital air corridors—low-altitude aerial lanes mapped by GPS, 5G networks, and AI coordination platforms—are being laid out to manage the safe, efficient routing of autonomous aerial vehicles above urban congestion.

To make this sky-based logistics system function seamlessly, regulatory bodies are harmonizing their rulesets to define new categories of pilot licensing, autonomous vehicle certification, and remote identification standards. What was once the purview of niche aviation lawyers is now the subject of national policy and international coordination, reflecting the urgent need for unified protocols in a rapidly expanding airspace.

In practice, drones are already proving themselves indispensable across multiple domains. In last-mile delivery, platforms like Amazon Prime Air, Alphabet's Wing, and Zipline are accelerating the movement of food, medications, and retail goods, particularly in high-density neighborhoods and hard-to-reach rural zones. Where traffic snarls or infrastructure gaps create delays on the ground, drones offer uninterrupted mobility overhead. These aerial couriers are not just faster—they're often greener, quieter, and more responsive to sudden spikes in demand or public need.

In the realm of infrastructure management, drones equipped with LiDAR, thermal imaging, and multispectral cameras are giving engineers the ability to map cities in ultra-high resolution. From inspecting bridges for microfractures to analyzing power lines for thermal anomalies, UAVs provide a layer of oversight that is safer, faster, and often more accurate than manual surveys. These data streams now feed directly into digital twin platforms—real-time 3D replicas of physical assets—where project managers and planners can simulate outcomes, monitor progress, and preempt failures.

Construction oversight has similarly been transformed. Project timelines can now be tracked day by day via autonomous drone flights, which document the evolution of work sites with centimeter-level precision. Anomalies or deviations from blueprint specifications are flagged early, reducing costly rework and enhancing coordination across subcontractors. The drone becomes not just a camera in the sky but a central node in the digitized construction lifecycle.

When disaster strikes—whether in the form of hurricanes, wildfires, or floods—drones have emerged as first responders of the air. They can be deployed within minutes to assess terrain changes, detect human heat signatures, and map hazardous zones, providing emergency teams with the situational awareness needed to coordinate relief and save lives. Their ability to fly in, collect critical information, and return autonomously has made them invaluable in fast-evolving scenarios where time is measured in lives.

At the heart of these capabilities is a leap forward in artificial intelligence and edge computing. Modern drones no longer require constant oversight or pre-scripted commands. They navigate independently, adjusting course to avoid obstacles, responding to changing weather conditions, and transmitting data in real time to cloud-based systems or localized command centers. As these machines grow smarter, faster, and more connected, they will underpin a new layer of urban intelligence—one where aerial mobility, infrastructure insight, and logistical fluidity converge to shape cities that are not just more efficient, but more resilient and responsive.

In this vision of the near future, drones are not merely flying devices. They are extensions of the city's nervous system—eyes, limbs, and logic engines in motion—quietly orchestrating the flow of goods, services, and information from above.

The future battlefield will not be defined solely by boots on the ground or jets in the sky — it will be shared by humans and machines, operating side by side, sometimes even autonomously. Artificial intelligence is transforming drones from reconnaissance platforms into fully autonomous weapons systems, forcing militaries around the world to rewrite doctrine in real time.

Already, AI-guided drones are assuming combat roles once thought to require human judgment. The United Kingdom's **StormShroud** drone is engineered to accompany manned aircraft into hostile airspace, jamming enemy radar systems and serving as an electronic escort. In Germany, **Stark's OWE-V** drone can execute semi-autonomous strikes up to 100 kilometers from its launch point — without operator input once airborne. Meanwhile, the United States Air Force is advancing its **Skyborg** program, a next-generation initiative to deploy AI-enabled "loyal wingmen" like the **XQ-58A Valkyrie** — drones capable of making independent combat decisions in complex, dynamic environments.

These technologies offer clear strategic advantages: they reduce the risk to human pilots, increase operational tempo, and enable persistent surveillance and strike capabilities far beyond traditional aircraft. But they also raise urgent ethical and legal questions. Who is accountable when an autonomous drone makes a mistake? What protocols are in place to prevent misidentification or unintentional escalation? Can machines be trusted with life-and-death decisions, or must the final call always remain human?

As lethal autonomous weapons systems (LAWS) evolve, the international legal framework lags behind. No global consensus exists on their regulation, and diplomacy continues to trail innovation. In this vacuum, nations race to develop capabilities that may one day outpace control.

Yet drones are not just instruments of war. They are also powerful tools for peace, stability, and environmental protection.

Across the globe, humanitarian organizations and UN peacekeepers are using drones to save lives, uphold human rights, and bring order to chaotic regions. In conflict zones, drones serve as unblinking observers — monitoring troop movements, surveying ceasefire areas, and documenting violations in places too dangerous for human presence. They offer transparency and accountability in missions where on-the-ground verification is impossible, helping peacekeepers identify emerging threats and act before violence reignites.

In refugee camps, drones are used to assess infrastructure needs — locating damaged shelters, blocked access roads, and sanitation failures. By capturing real-time imagery over wide areas, they give aid organizations a bird's-eye view of conditions on the ground, speeding up the delivery of food, medicine, and water.

In the aftermath of natural disasters, drones are now among the first responders. From **Nepal to Haiti**, UAVs have transformed how rescue teams operate in the critical hours following a catastrophe. Aerial surveys pinpoint survivors amid collapsed buildings, track blocked roads, and identify areas cut off by floodwaters or landslides.

Medical drones drop essential supplies — from blood packets to antibiotics — into regions that are otherwise unreachable. In areas where communication towers have been destroyed, drones can restore temporary **Wi-Fi or cellular service**, allowing communities to contact loved ones and rescuers to coordinate their efforts.

In moments where every minute matters, drones offer speed, agility, and endurance that human crews alone cannot match.

Far from the chaos of war or disaster, drones are quietly reshaping how we study and protect the planet. In the battle against climate change, they serve as unmatched data collectors — capturing images and metrics from the remotest corners of the earth.

Drones monitor **glacial retreat** and **coastal erosion**, track **wildlife populations** using thermal sensors, and measure **air and soil quality** in polluted urban zones. In the rainforests of South America and Southeast Asia, they help scientists detect **illegal logging** and map **deforestation** with unprecedented clarity.

What once required helicopters or costly satellite imagery can now be done with compact UAVs at a fraction of the cost — and with greater frequency and precision. These flying laboratories are generating the most detailed, real-time environmental datasets humanity has ever had.

Whether in conflict or in conservation, drones are redefining how we see, understand, and intervene in the world. As they continue to evolve, they present both a profound opportunity and a profound responsibility — to wield them wisely, ethically, and for the betterment of life on Earth.

The next decade will not just see drones in the sky—it will see drones embedded in society, guiding ambulances, inspecting roads, delivering meals, and perhaps even flying families across a city. From swarms to singular superdrones, the convergence of AI, autonomy, and aerial mobility is reshaping everything from warfare to wildlife management.

But the future of drones is not just about what they can do—it is about what we allow them to do. The technologies are ready. The skies are opening. Now comes the challenge of vision, governance, and global coordination.

Because in the unmanned future, the sky is no limit—it's just the beginning.

Chapter 11: Shadows Over the Base – Drone Incursions at Military Installations

In recent years, a series of mysterious and unauthorized drone sightings over sensitive military installations has sparked alarm among defense officials, intelligence agencies, and lawmakers alike. These incidents—ranging from single drone flyovers to coordinated swarms—have raised urgent questions about national security vulnerabilities, foreign surveillance, and the limits of current countermeasures.

As drones become cheaper, smaller, and more capable, military bases, missile silos, nuclear sites, and naval assets are increasingly finding themselves under uninvited aerial observation. The skies above secure facilities are no longer a sanctuary—they are contested territory.

As unmanned aerial vehicles become increasingly accessible and sophisticated, the threat of their misuse has grown just as rapidly. Drones are no longer limited to hobbyists and commercial enterprises — they are now tools for espionage, drug trafficking, unauthorized surveillance, terrorist attacks, and battlefield reconnaissance. In response, nations and institutions have developed an entirely new class of defense technology: counter-unmanned aircraft systems, or C-UAS.

These systems are designed to detect, track, identify, and neutralize unauthorized or hostile drones in a wide range of environments — from military bases to critical infrastructure and crowded public spaces. The foundation of this technology lies in real-time awareness. By leveraging deep learning, computer vision, and multi-sensor fusion, modern C-UAS platforms can detect UAVs using standard and specialized cameras, track their movements across a layered network of sensors, and even identify specific drone models using advanced re-identification algorithms that work across changing angles, lighting, and terrain.

Among the most advanced systems in operation today is the Aaronia AARTOS, a modular and scalable C-UAS solution deployed at sensitive sites including international airports. It provides continuous surveillance, threat prioritization, and automated geofencing — monitoring the skies for anomalies and unauthorized incursions with remarkable precision.

Once a UAV is identified as a threat, the next challenge is neutralization. Countermeasures generally fall into two broad categories: kinetic and non-kinetic. Kinetic solutions are physically destructive and include anti-drone interceptors — drones that hunt and disable intruding UAVs — as well as net launchers, projectiles, and even repurposed anti-aircraft systems like those adapted within Israel's Iron Dome defense architecture. High-energy laser systems, once the domain of science fiction, are now being tested and deployed to silently burn drones out of the sky in seconds.

Non-kinetic methods are subtler but no less effective. These include radio frequency jamming, which disables a drone's communication with its operator; GPS spoofing, which deceives its navigation system; and directed microwave attacks that overload onboard electronics. In more sophisticated operations, cyber teams can hijack a drone's command-and-control link, seizing control mid-flight and landing it safely or rerouting it entirely.

Portable jamming units are becoming commonplace — from U.S. border patrol deployments to Italy's 17th Anti-Aircraft Artillery Regiment "Sforzesca," which uses handheld jammers to secure public events, national landmarks, and transportation hubs. The future of drone defense may lie in the very systems it seeks to defeat: intelligent, autonomous drone swarms capable of intercepting hostile UAVs in real time through coordinated flight and decision-making. These C-UAS swarms represent a vision of airspace security that is as dynamic and adaptable as the threats they are designed to counter.

As drone capabilities continue to evolve, so too must our defenses — not only in military contexts, but across every sector where security, privacy, and public safety intersect with the open sky.

As drone activity surges worldwide, regulatory frameworks are rapidly evolving to integrate UAVs safely into civil and military airspace. National aviation authorities are developing Unmanned Traffic Management (UTM) systems to coordinate drone flights with manned aviation, especially in urban and commercial settings.

The International Civil Aviation Organization (ICAO) began examining UAV regulation as early as 2005, culminating in a 2011 foundational report. France was among the first nations to implement a national regulatory framework based on ICAO guidance. Soon after, the

U.S. Federal Aviation Administration (FAA) and the European Union Aviation Safety Agency (EASA) began building comprehensive legal and safety structures for drone use.

In 2021, the Federal Aviation Administration (FAA) implemented the **Remote Identification (Remote ID) Rule**, a regulatory milestone aimed at bringing greater accountability and transparency to the growing drone ecosystem. Under this rule, all commercial drones — as well as any recreational drone weighing more than 250 grams — are required to broadcast identifying information during flight. This includes the drone's real-time location and altitude, its point of takeoff, and the identity of the registered operator. Designed to function like a digital license plate for the sky, Remote ID enables authorities to track and monitor drones in shared airspace, enhancing both air traffic safety and public trust in unmanned aviation.This rule, designed to improve airspace transparency and public safety, is currently under judicial review in RaceDayQuads v. FAA, a lawsuit challenging privacy and procedural concerns.

Europe has adopted a complementary system that includes the Class Identification Label, which certifies that a UAV meets safety and technical criteria for its operational category (C0–C6). These classifications determine where and how a drone may fly, and whether pilot training or registration is required.

The proliferation of advanced UAVs has raised geopolitical and security concerns. To curb the unchecked spread of powerful drone technologies, many countries adhere to the Missile Technology Control Regime (MTCR), an international agreement that restricts the export of systems capable of delivering a 500 kg payload over 300 kilometers.

UAVs that meet or exceed these thresholds — especially armed or long-range reconnaissance drones — are subject to strict licensing, end-user vetting, and bilateral agreements. The MTCR aims to prevent the weaponization of drones by non-state actors or hostile regimes, while balancing the legitimate needs of defense partners and civilian industries.

Between 2019 and 2021, a series of perplexing and widely publicized drone encounters unfolded off the coast of Southern California, involving multiple U.S. Navy vessels operating in restricted waters. Over several consecutive nights, destroyers such as the **USS Kidd** and **USS Rafael**

Peralta reported being swarmed by unidentified aerial vehicles exhibiting capabilities well beyond those of ordinary consumer drones. These UAVs conducted high-speed, coordinated maneuvers — often in poor weather and strong winds — and remained airborne for durations that far exceeded the limitations of standard battery-powered systems. The pattern of repeated activity suggested deliberate, externally coordinated operations.

The incidents triggered joint investigations by the U.S. Navy, Coast Guard, FBI, and Department of Defense, yet despite their efforts, no definitive source or operator was ever publicly identified. The mystery fueled speculation across the defense community and media, with theories ranging from foreign surveillance operations by adversaries such as China or Russia, to clandestine testing of commercial or modified drone platforms, or even undisclosed U.S. military or intelligence exercises.

Regardless of origin, what the events exposed was sobering: even some of the world's most technologically advanced and heavily defended warships were vulnerable to low-cost, hard-to-attribute drone incursions. The Navy acknowledged significant gaps in detection, tracking, and attribution — a stark reminder that in the age of asymmetric threats, sophisticated military assets can be challenged by cheap, elusive systems that exploit the blind spots of traditional defense infrastructure.

The threat posed by unidentified drones is not confined to naval operations or open-airspace surveillance. In recent years, drones have been reported operating above some of the most sensitive facilities in the United States — nuclear power plants and intercontinental ballistic missile (ICBM) fields — raising serious concerns about critical infrastructure vulnerability and national defense.

In 2014, a wave of drone sightings over nuclear reactors in France prompted a sweeping national security review, highlighting how even tightly secured energy facilities could be surveilled or threatened from above with little warning. Five years later, a similar event occurred on U.S. soil when drones were spotted flying over the **Palo Verde Nuclear Generating Station** in Arizona — the largest nuclear plant in the country. The drones evaded capture, and their operators were never identified, igniting fears of possible reconnaissance, dry-run sabotage attempts, or cyber-physical threat scenarios.

The concerns intensified in 2021, when reports surfaced of drone activity near **Minuteman III missile silos** in **Montana and Wyoming**. Witnesses described precise flight paths and persistent tracking over silo fields — behavior inconsistent with recreational use. These incidents appeared to involve coordinated operations and were taken seriously by military and intelligence agencies, though official findings remained undisclosed.

What these events revealed was not merely the audacity of the incursions, but the inadequacy of current systems to prevent or respond to them. Federal authorities lacked a unified protocol for drone incursions over high-value assets, and confusion persisted among law enforcement, military units, and federal regulators regarding jurisdiction and the legal use of countermeasures. Moreover, some analysts speculated that these events could represent probing operations — attempts by adversaries to map U.S. response times, radar coverage gaps, or the electromagnetic signature of critical infrastructure.

In the age of unmanned systems, these high-profile intrusions underscore a sobering reality: airspace sovereignty is no longer guaranteed, even over the most fortified domains of national security.

One of the most confounding challenges in responding to drone incursions is the question of attribution. Unlike manned aircraft or traditional ballistic threats, small UAVs are often designed to elude identification. They typically lack transponders, fly at low altitudes beneath radar thresholds, and exhibit erratic flight paths that frustrate pattern recognition and predictive tracking. This ambiguity makes it difficult to determine not only who is operating the drone, but what their intent may be.

Potential actors span a wide spectrum. Foreign intelligence agencies may deploy drones to gather photographic surveillance or intercept electronic communications. Civilian hobbyists — either unwitting or defiant — can wander into restricted airspace with commercially available drones, unaware of the potential consequences. More troubling still are the scenarios involving domestic adversaries, including extremist or terrorist groups, who may be using drone incursions as a form of reconnaissance, probing for vulnerabilities in U.S. defenses or critical infrastructure.

Even when a drone is successfully recovered, identifying its origin remains a daunting task. Many are assembled using off-the-shelf components, making it nearly impossible to trace them to a specific manufacturer or operator. Others are intentionally obfuscated — stripped of serial numbers, outfitted with encrypted communications systems, or constructed using foreign parts that further muddy the investigative waters. In several recent cases, recovered drones were found to contain Chinese-made electronics or secure modules, intensifying suspicions of foreign state involvement and sparking renewed debate over the supply chain security of drone technologies.

In this shadowy airspace, where threat and anonymity coexist, attribution remains the missing link — a gap that leaves responders navigating not only the drone's flight path, but a fog of geopolitical uncertainty.

In response to the rising threat posed by unmanned aerial vehicles, the Pentagon has rapidly expanded its arsenal of Counter-Unmanned Aerial Systems (C-UAS). What began as a niche concern has evolved into a central pillar of defense strategy, as drones — once the domain of hobbyists and reconnaissance missions — now represent a potential vector for espionage, disruption, and direct attack.

The Department of Defense has deployed a range of technologies to confront this growing menace. Electronic warfare jammers, such as the **DroneDefender** and **DroneGun Tactical**, disrupt drone communications by targeting their radio frequency (RF) links, forcing them to land or return to base. **Directed energy weapons** — including high-powered microwave arrays and solid-state lasers — are under active testing and deployment, particularly for scenarios involving drone swarms, where traditional kinetic responses may be too slow or resource-intensive. Meanwhile, **kinetic interceptors** remain a vital layer of last-resort defense, encompassing everything from net-firing systems and interceptor drones to repurposed anti-aircraft artillery adapted for low-altitude engagements.

Complementing these active defenses are **passive detection networks**, which rely on an integrated mesh of acoustic sensors, RF scanners, and optical surveillance tools to map the flight paths of intruding drones and anticipate swarm patterns. These systems not only improve early warning capabilities but also help refine the understanding of airspace vulnerabilities across different environments.

To coordinate and standardize these efforts, the U.S. Department of Defense has established the **Joint C-UAS Office (JCO)** — a central body tasked with ensuring interoperability, doctrine development, and strategic deployment of counter-drone systems across all service branches.

Yet, significant challenges remain. Rules of engagement (ROE) for drones are often ambiguous, particularly when the aircraft in question is not actively engaging in a hostile act. In civil airspace, especially near populated areas, the use of kinetic or electronic countermeasures may conflict with Federal Aviation Administration (FAA) regulations and public safety norms. Furthermore, the very sensors designed to detect drones can be overwhelmed in crowded RF or urban environments, where **false positives**, clutter, and environmental noise degrade situational awareness and raise the risk of misidentification.

As the drone threat continues to evolve, so too must the legal, technological, and operational frameworks designed to contain it — or risk leaving critical airspace exposed to an increasingly invisible and unpredictable adversary.

Congress has grown increasingly engaged with the escalating challenge of drone incursions, recognizing the national security, privacy, and safety implications posed by unmanned aerial systems operating in or near sensitive airspace. In recent years, a series of bipartisan legislative efforts has sought to address the technological, jurisdictional, and regulatory gaps that allow hostile or unauthorized drones to operate with relative impunity.

One of the most prominent proposals, the **Drone Origin Security Enhancement Act**, aims to restrict federal agencies from procuring foreign-manufactured drones — particularly those produced by Chinese firms — out of concern for data security and potential espionage. This measure reflects a broader concern within the intelligence community that foreign-made UAVs could act as covert surveillance platforms, feeding information back to adversarial governments through encrypted or hidden channels.

Other legislative initiatives have focused on expanding the authority of domestic security agencies. Proposed counter-UAS legislation would grant the **Department of Homeland Security (DHS)** and the **Department of Defense (DoD)** broader powers to interdict, disable, or neutralize drones operating near critical infrastructure, such as military bases, nuclear plants, and airports. These efforts gained traction following high-profile incursions that exposed weaknesses in both detection and response.

The **National Defense Authorization Act (NDAA) for 2023** included dedicated funding for the expansion of drone detection systems at key defense installations, including ICBM fields and strategic launch facilities. This allocation signaled congressional recognition that the threat landscape has outpaced existing surveillance capabilities — and that the skies above military assets can no longer be considered secure by default.

Lawmakers are also pushing for the establishment of a **national drone incident reporting system** to track and catalog unauthorized UAV activity across jurisdictions. Alongside this initiative are calls for **real-time coordination protocols** between civilian air traffic control and military installations, ensuring that anomalies are quickly assessed and responded to without delay or miscommunication.

Finally, legislation has proposed the development of **standardized no-fly zones** reinforced by mandatory **geofencing technology** — digital barriers that prevent drones from operating in protected areas. These zones, if broadly adopted and enforced, could serve as a first line of defense against casual violations and provide a legal and technological framework for intercepting more malicious threats.

As the drone frontier expands, congressional action will play a pivotal role in determining whether U.S. airspace remains safe, sovereign, and secure — or becomes a battleground of unregulated activity and delayed response.

The pattern of drone incursions—particularly those involving unknown operators and sensitive installations—has led many experts to conclude that these events are not merely the work of curious hobbyists or accidental airspace violations. Rather, the persistence, coordination, and tactical nature of many of these intrusions point to a more calculated objective.

Defense analysts increasingly warn that low-level, repeated drone activity near critical sites may be part of a broader strategy to **map radar blind spots**, **test electronic countermeasure readiness**, or **gather optical and signals intelligence** from high-value military and infrastructure targets. The drones involved are often small, quiet, and inexpensive, yet their movements suggest purpose and training, not randomness.

Such operations are consistent with the doctrine of **gray zone warfare**—a form of conflict waged below the threshold of open hostilities, where attribution is difficult and escalation is deliberately avoided. In this murky space, adversaries employ deniable technologies to probe defenses, disrupt routines, and shape perceptions without triggering conventional military responses.

Drones, by their very design, are ideal tools for this kind of strategic manipulation. They blur the line between surveillance and provocation, between reconnaissance and sabotage. In the wrong hands, they are not simply nuisances or novelties—they are instruments of stealth, deception, and influence, capable of undermining security not through brute force, but through ambiguity and asymmetry.

As drones proliferate, the once-secure airspace above military installations has grown increasingly vulnerable. What was once protected by altitude, restricted zones, and perimeter defenses is now accessible to small, commercially available aircraft capable of slipping unnoticed through gaps in radar coverage. The next major breach of national security may not come in the form of a stealth bomber or hypersonic missile, but rather a handful of plastic-bodied UAVs — quiet, low-flying, and invisible to conventional defenses — gathering intelligence, probing weaknesses, or delivering payloads in ways the architects of legacy systems never imagined.

In this new threat environment, defending strategic sites demands more than vigilance — it demands transformation. Continuous detection and tracking systems must be deployed across all critical infrastructure, from missile silos to naval bases and command centers. Unified response protocols must be established that allow for rapid escalation — from passive monitoring to electronic jamming, to physical interdiction — without delay or confusion. Legal authority must be clear and actionable, empowering defenders to engage airborne threats with precision, speed, and without fear of jurisdictional ambiguity. And above all, national defense must evolve in its assumptions: recognizing that surveillance and reconnaissance are no longer the exclusive domain of satellites or state-sponsored spy planes. They now extend to off-the-shelf drones flown from parking lots, treelines, or the back of a van.

The battlefield is no longer distant. It may be hovering just above the fence line — silent, circling, and watching.

Chapter 12: Drones Over Donbas – Unmanned Warfare in the Russia-Ukraine War

When Russia launched its full-scale invasion of Ukraine in February 2022, the world witnessed not only the return of large-scale conventional warfare to Europe—but the dawn of drone warfare as a central feature of modern conflict. What had previously been limited to targeted strikes and surveillance in asymmetric conflicts now erupted into drone-on-drone engagements, swarm tactics, kamikaze strikes, and mass battlefield surveillance—all in one of the most technologically documented wars in history.

At the outset of the war in Ukraine, the **Turkish-made Bayraktar TB2** quickly emerged as a symbol of resistance and battlefield ingenuity. Developed by **Baykar**, the TB2 is a medium-altitude, long-endurance (MALE) drone equipped with precision-guided munitions and advanced electro-optical and infrared targeting systems. Its early battlefield successes — striking Russian convoys, artillery batteries, and fuel columns — captured both strategic advantage and public imagination, earning the drone near-mythic status among Ukrainians and their allies.

The TB2 played a pivotal role in the defense of **Kyiv**, **Kharkiv**, and **Mykolaiv**, coordinating attacks on advancing Russian armor and softening enemy positions in contested zones. It was credited with the destruction of multiple **Buk** and **Pantsir** air defense systems — formidable targets that typically required high-risk manned sorties to engage. One of the most dramatic episodes tied to the TB2 was its reported role in targeting the **Moskva**, the flagship of the Russian Black Sea Fleet, prior to its sinking — a symbolic and tactical blow to Russian naval power in the region.

What the TB2 demonstrated was not merely tactical success, but a new model of airpower: **cost-effective, precise, and expendable**. Where traditional aircraft require hundreds of millions in development and support, the TB2 delivered measurable strategic effects at a fraction of the cost, with far lower risk to personnel. It became a case study in how relatively modest drone platforms, when used creatively and integrated into flexible command structures, can reshape battlefield dynamics.

As the war progressed, however, Russian forces adapted. Improved electronic warfare, layered air defenses, and tighter operational security made strike missions more perilous for the TB2. In response, Ukrainian forces shifted the drone's role from direct attack to **reconnaissance and decoy operations**, highlighting the fluid nature of drone doctrine under real-time battlefield pressure. The TB2's evolution in the theater of war became a microcosm of drone warfare itself — fast-moving, adaptive, and redefining the balance between manned and unmanned force projection.

In response to Ukraine's high-profile success with the Bayraktar TB2, Russia moved swiftly to expand and refine its own drone capabilities, leaning heavily on **loitering munitions** — a class of drones designed to hover over a battlefield until a target is identified, then strike with lethal precision.

Foremost among these was the **ZALA Lancet**, a series of kamikaze drones developed by Russia's Kalashnikov-affiliated ZALA Aero. Compact, efficient, and deadly, the Lancet emerged as a precision-guided weapon system tailored for asymmetric advantage. These drones proved particularly effective in disabling **NATO-supplied artillery platforms** — including **M777 howitzers** and **Caesar self-propelled guns** — as well as Ukrainian **air**

defense systems and **command-and-control vehicles**. The **Lancet-3**, often launched from rail-based or mobile ground platforms, became a mainstay of Russian frontline operations. It was frequently deployed in **hunter-killer formations** alongside surveillance drones like the **Orlan-10**, which would first locate and track high-value targets before directing the Lancet to engage.

However, it was Russia's partnership with Iran that marked one of the most controversial and consequential shifts in drone warfare during the conflict. In late 2022, Russia began acquiring **Shahed-136 drones** — long-range, GPS-guided loitering munitions with high-explosive payloads. Rebranded as **Geran-2** by Russian forces, these drones became central to Moscow's winter campaign of attritional warfare. From **October 2022 onward**, swarms of Shaheds were launched in mass barrages aimed at Ukraine's **energy infrastructure**, plunging cities into darkness and cold as winter took hold. They also served as instruments of **psychological warfare**, striking urban centers, civilian targets, and logistical nodes — sometimes without warning, often in coordinated waves.

The drones' low flight paths and swarm tactics posed serious detection and interception challenges for Ukrainian defenses. While the Shaheds themselves were relatively inexpensive — estimated at just tens of thousands of dollars per unit — the cost of intercepting them was orders of magnitude higher. Ukrainian forces were often forced to use **advanced surface-to-air missiles** to bring down the drones, creating a stark economic imbalance that underscored the asymmetry at the heart of modern drone warfare.

Russia's integration of the Lancet and Shahed systems marked a shift in doctrine — one in which unmanned systems were no longer supplementary tools but **frontline instruments of attrition, suppression, and terror**. The ability to deploy mass-produced drones for both tactical and strategic impact demonstrated how accessible, adaptable, and politically deniable drone warfare had become — changing not only the contours of the Ukraine conflict, but the global calculus for military planners facing similar threats.

As the war progressed, both Ukraine and Russia moved swiftly to adapt **civilian drone platforms** for frontline combat — a phenomenon that came to define a new era of low-cost, high-impact warfare. **DJI quadcopters**, originally designed for aerial photography, and **first-person view (FPV) racing drones**, favored by hobbyists, were transformed into lethal battlefield assets. Modified FPV drones, fitted with shaped charges, were manually piloted into enemy armor, trenches, and fortified bunkers, effectively becoming guided anti-personnel and anti-vehicle weapons. Civilian quadcopters were retrofitted with custom release mechanisms, allowing troops to drop grenades or mortar rounds with remarkable precision onto targets often hidden from direct line of sight.

This wave of improvisation was not centrally directed but organically grown. **Crowdfunding campaigns**, civilian volunteer groups, and grassroots engineering networks across Ukraine — and to a lesser extent in Russia — fueled the rapid acquisition, modification, and repair of drones on the battlefield. In a matter of months, small infantry squads gained the kind of **intelligence, surveillance, and reconnaissance (ISR)** and strike capability once reserved for major state actors with advanced air forces. The democratization of drone warfare reshaped tactical doctrine, placing aerial situational awareness and precision strike capabilities into the hands of ordinary soldiers.

But as drone use surged, so did efforts to neutralize them. The **electromagnetic spectrum** — once a silent backdrop to military operations — became a contested frontline in its own right. Russia deployed sophisticated **electronic warfare (EW)** platforms, including the **Krasukha-4** and **Leer-3**, to jam GPS signals, interfere with communications links, and spoof drone control systems. These tools disrupted both autonomous and manually piloted UAV operations, forcing a constant adaptation cycle.

In response, Ukraine — with critical Western support — employed **anti-drone rifles**, **mobile jamming units**, and **counter-UAS radar systems** to blunt Russian drone operations. What emerged was a kind of **technological arms race in miniature**, an evolutionary loop driven by field experience: new drones led to new countermeasures, which in turn spurred the creation of new drone designs, tactics, and frequencies to evade them.

Drone losses on both sides numbered in the thousands. Yet commanders tolerated — even embraced — this level of attrition, recognizing that the tactical advantage conferred by persistent aerial ISR and precision strikes outweighed the material cost. The battlefield had entered a new phase, one not of shock-and-awe from above, but of **continuous innovation at low**

altitude — where the skies buzzed with cheap, lethal ingenuity, and every advance was met with a countermeasure, every setback with a workaround.

The Russia-Ukraine war has fundamentally redefined how militaries around the world think about the modern battlefield. What was once theoretical — the role of drones in high-intensity conflict — has now become tactical doctrine. One of the clearest lessons to emerge is that **swarms and loitering munitions** are no longer peripheral technologies; they are central to **combined arms operations**, shaping engagements across every domain from artillery coordination to air defense suppression.

Drones, whether custom-built or improvised from consumer platforms, are treated not as prized assets to be conserved, but as **expendable tools of warfare**. They are reused when possible, sacrificed when necessary, and continuously replaced as the tempo of conflict demands. The low cost of production and the relative ease of deployment have allowed commanders to rethink risk — pushing UAVs into missions that would be unthinkable for manned aircraft.

Crucially, the war demonstrated that **air superiority is no longer a prerequisite for aerial advantage**. Even in contested skies filled with radar threats and jamming systems, drones have carved out a persistent presence, operating below detection thresholds, beyond visual line of sight, and often outside conventional air defense paradigms. This has enabled both sides to field improvised aerial capabilities under conditions that would ground traditional aircraft.

The conflict also revealed how rapidly **civilian drones can be militarized**, spawning what are effectively "**drone militias**" supported by crowdfunding, volunteer engineering groups, and local battlefield workshops. From quadcopters to FPV racing drones, the ability to adapt off-the-shelf platforms into precision weapons has empowered irregular forces and blurred the line between state-sponsored military and grassroots innovation.

But perhaps the most transformative realization is this: in modern warfare, **information dominance** is as critical as firepower. The side that sees more, reacts faster, and targets with greater precision gains a decisive edge. Drones have become the eyes of the battlefield — extending the reach of every rifle, every artillery piece, and every command decision. In the evolving geometry of conflict, the high ground is no longer measured in elevation — but in **real-time situational awareness**, delivered from above.

Chapter 13: Eyes of the Crescent – Drone Warfare in the Middle East

The Middle East has become one of the most active testing grounds for drone warfare. Unlike previous decades of conventional conflict, modern engagements in this region feature a complex mix of state actors, militias, and proxy forces—all employing unmanned aerial systems for reconnaissance, targeted strikes, and psychological warfare.

From Iranian-manufactured kamikaze drones to Israeli deep-strike UAVs and Turkey's rising drone diplomacy, the Middle East now showcases the full spectrum of drone capabilities in hybrid conflict.

Over the past decade, **Iran has quietly emerged as one of the most prolific drone powers in the world**, building an expansive arsenal of unmanned aerial systems designed for **asymmetric warfare** and exporting them to a growing network of aligned militias and proxy forces. What began as a modest domestic program has evolved into a transnational drone doctrine — one that allows Iran to project power across the Middle East without direct confrontation, and often without attribution.

At the heart of Iran's UAV fleet are several key platforms. The **Shahed-129**, modeled in part on the American MQ-1 Predator, is a long-endurance surveillance and strike drone capable of operating at extended ranges with real-time targeting capability. The **Mohajer-6**, a smaller tactical UAV operated by the Islamic Revolutionary Guard Corps (IRGC), has seen widespread use in regional conflicts and has become a staple in Iran's export strategy to partner forces. Perhaps most infamous is the **Shahed-136** — known in Russian service as the **Geran-2** — a low-cost, GPS-guided loitering munition deployed in Ukraine, Syria, Iraq, and Yemen. Launched in swarms, these "kamikaze drones" have proven difficult to detect and costly to intercept.

Iran's drone doctrine is built not only on indigenous development but on **strategic proliferation**. In Lebanon, **Hezbollah** has used Iranian drones to conduct reconnaissance missions deep into Israeli territory and to probe air defense systems with mock or decoy flights. In Yemen, **Houthi rebels** have executed drone attacks on Saudi oil fields, airports, and population centers

— striking targets hundreds of miles from launch points with precision previously unavailable to non-state actors. In Iraq, Iranian-aligned militias have targeted **U.S. and coalition bases** with one-way attack drones, often operating from concealed, mobile launch platforms.

These UAVs are frequently deployed from **civilian vehicles, remote deserts, or improvised sites**, making them elusive and difficult to preempt. Their effectiveness lies not in technological superiority, but in **volume, saturation, and unpredictability**. Individually, they may be slow, loud, and unsophisticated — but collectively, they overwhelm traditional air defense systems that were never designed to counter dozens of small, low-flying threats at once.

By enabling proxy forces to strike at state-level adversaries without triggering direct retaliation, Iran has redefined the utility of drones in regional conflict. Its model of **plausibly deniable, distributed airpower** has become a case study in how unmanned systems can shift the balance of power — not through brute force, but through persistent pressure and tactical ambiguity.

Hezbollah, long known for its ground-based guerrilla tactics, has evolved into a **hybrid force** capable of integrating air reconnaissance and unmanned strike capability into its operational doctrine. With the support of Iran, the group has transformed from a regional militia into a technologically equipped actor capable of projecting power beyond its traditional borders.

In **2022**, Hezbollah escalated tensions by launching drones over **Israel's Karish gas platform** in the eastern Mediterranean, a strategically sensitive site central to regional energy production. Although intercepted, the incident sparked a flurry of diplomatic and military responses, highlighting how unmanned platforms can be used not only for surveillance or strike, but for **strategic signaling** and geopolitical pressure.

Hezbollah has also employed tactical drones in battlefield operations, particularly for **artillery spotting and post-strike assessment**, increasing the accuracy and effectiveness of indirect fire against Israeli positions. In other cases, drones have been used for **propaganda purposes**—recording attacks on Israeli outposts and broadcasting them to bolster internal morale and demonstrate operational reach.

The UAVs used by Hezbollah are widely believed to be **Iranian-supplied**, often based on models like the Ababil or Mohajer series, though many are **locally modified** for specific mission profiles. These drones now function as **force multipliers** in asymmetric engagements, giving the group real-time eyes in the sky and the ability to strike selectively without exposing its fighters to direct risk.

By blurring the line between **surveillance and strike**, Hezbollah's drone operations exemplify a growing trend in modern conflict: the integration of low-cost, high-impact technologies into non-state military campaigns. In doing so, the group has redefined what is possible for irregular forces, challenging traditional military hierarchies with tools once reserved for state powers. Israel has long been at the forefront of unmanned aerial warfare, pioneering many of the doctrines and technologies that now define drone combat around the world. As early as the **1982 Lebanon War**, Israel deployed unmanned aerial vehicles (UAVs) for real-time battlefield surveillance and decoy missions—decades before most militaries even considered drones viable combat assets. Since then, the **Israeli Air Force (IAF)** has developed one of the most advanced and battle-tested drone fleets on the planet.

At the core of Israel's UAV arsenal is the **IAI Heron**, a long-endurance surveillance platform used extensively for border patrol, reconnaissance, and persistent battlefield observation. Complementing it is the **Hermes 900**, a versatile multi-role drone capable of performing **strike missions**, **signals intelligence (SIGINT)** collection, and **target acquisition** with high precision. Perhaps most uniquely Israeli is the **Harop**, a loitering munition designed to autonomously detect and destroy enemy radar installations. The Harop is not just a drone—it is a **flying hunter-killer**, merging surveillance and firepower into a single expendable asset.

During repeated conflicts in **Gaza** and **Syria**, Israel has employed its drones with surgical precision. UAVs have been used for **targeted assassinations** of senior figures in Hamas and Palestinian Islamic Jihad (PIJ), striking with minimal warning and maximum effect. In some operations, drones have carried out so-called **"knock on the roof"** strikes—non-lethal warning shots intended to prompt evacuation of a building before a larger munition is deployed. This tactic reflects Israel's effort to reduce civilian casualties while maintaining operational effectiveness.

In the electromagnetic domain, Israeli drones have also played a critical role. Prior to manned airstrikes, UAVs have been deployed to **jam enemy communications**, disrupt command-and-control channels, and blind adversary defenses—creating a seamless integration between intelligence gathering, electronic warfare, and kinetic engagement.

Israel's UAV doctrine is characterized by **tight integration between real-time surveillance, precision targeting, and automated response cycles**. Rather than operating drones as isolated tools, the IAF uses them as **core nodes in a networked warfighting system**, linking battlefield data to actionable decisions within seconds. This fusion of intelligence and lethality has made Israel's drone strategy a **global benchmark**, studied by allies and adversaries alike.

In recent years, **Turkey has rapidly emerged as a leading drone power**, leveraging indigenous innovation, strategic export diplomacy, and battlefield experience. At the center of this ascent is the **Bayraktar TB2**, a medium-altitude, long-endurance UAV developed by **Baykar**. Combat-tested in multiple theaters—**Syria, Libya, Nagorno-Karabakh, and Ukraine**—the TB2 has demonstrated remarkable versatility and reliability, becoming one of the most widely used armed drones outside NATO inventories. It has also been deployed extensively along **Turkey's**

southern border, where drone warfare plays a central role in counterterrorism and border security operations.

Against **Kurdish militant groups** such as the PKK and YPG in **Iraq and Syria**, Turkish drones have been used to provide **intelligence, surveillance, and reconnaissance (ISR)** for special operations forces, conduct **precision strikes** against high-value targets, and monitor **cross-border smuggling and infiltration routes**. These missions are often conducted unilaterally and without the approval of U.S.-controlled airspace—a posture that has led to **diplomatic tensions**, especially within NATO, but also underscored Turkey's growing independence and technological self-sufficiency in unmanned systems.

The **United States**, long the global pioneer of drone strike warfare, first institutionalized UAV combat operations in the Middle East during the **post-9/11 Global War on Terror**. While much of the Pentagon's strategic focus has since shifted toward the Asia-Pacific region, U.S. drones remain actively engaged in the Middle East. **MQ-9 Reapers** continue to patrol over **Iraq and Syria**, providing ISR and executing targeted strikes against **ISIS remnants** and **Iranian-backed militias**. These drones also support **Kurdish partner forces**, despite the increasingly delicate regional politics surrounding their deployment.

The evolving threat landscape has prompted the U.S. military to fortify its drone defenses across the region. **Counter-drone systems**—capable of detecting, jamming, or destroying hostile UAVs—have been deployed at bases in **Al-Tanf, Erbil, and Baghdad**, where **drone swarm attacks** have become an emerging threat. While the United States retains **technological superiority**, the proliferation of low-cost, armed UAVs among adversaries has forced a doctrinal shift in **base defense, convoy protection, and ISR redundancy**. Drone dominance can no longer be assumed—it must be actively defended.

In more complex and densely populated environments like **Gaza**, drone use presents acute ethical and operational challenges. The Israeli Defense Forces (IDF) have integrated drones into **urban warfare**, using miniature quadcopters and tactical UAVs to locate and strike **tunnel entrances** and weapons caches embedded in civilian infrastructure. At the same time, **militant groups** have begun co-opting commercial drones for surveillance, psychological operations, and harassment—blurring the line between state and non-state drone operators.

The IDF's strategy of using drones for **"over-the-shoulder" close air support**—in which UAVs accompany infantry through urban corridors and provide near-instant precision fire—represents a new paradigm of **micro-scale air dominance**. But these tactics, while effective, are subject to global scrutiny. **The use of drones in urban combat zones**, where enemies exploit civilian cover and conduct operations from residential buildings, demands a level of precision and accountability that stretches the limits of current drone doctrine. In such environments, every strike is not just a military action—it is also a test of **discrimination, proportionality, and the legitimacy of force** in modern conflict.

The **Middle East remains both a laboratory and a battleground for drone warfare**, a region where the skies are increasingly populated not by traditional air forces alone, but by a diverse and expanding fleet of unmanned systems. From **Gaza's rooftops to the Syrian desert**, the airspace hums with the constant activity of UAVs — some silently scanning the terrain, others delivering munitions with precision, and many transmitting live video feeds to operators watching from command centers thousands of miles away.

In this fragmented and volatile theater, drones serve **multiple roles simultaneously**. They are **strategic instruments of statecraft**, extending the reach of nations like Iran, Israel, Turkey, and the United States without requiring boots on the ground. They are **tactical force multipliers** for non-state actors and militias, providing intelligence, surveillance, and the ability to strike targets once considered unreachable. And they are **psychological weapons**, wielded to coerce populations, spread propaganda, and unsettle adversaries through their very presence and unpredictability.

As technologies in **artificial intelligence, autonomous swarming, and low-observable drone design** continue to advance, the region is poised to enter an even more complex phase of unmanned warfare. The next generation of drones may be stealthier, faster, and more independent — capable of executing missions without real-time human oversight, and in many cases, without attribution. In such a landscape, deniability becomes a strategic asset, and **the balance of power can shift silently, invisibly — one drone at a time**.

The U.S. Air Force's CCA program aims to develop autonomous unmanned aerial vehicles (UAVs) that can operate alongside crewed fighter jets, enhancing mission capabilities through manned-unmanned teaming (MUM-T). These UAVs are designed to perform a variety of roles, including surveillance, electronic warfare, and strike missions, while being more cost-effective and expendable than traditional manned aircraft.

The YFQ-44A "Fury", a next-generation autonomous aircraft under development by Anduril Industries, represents a bold step forward in the fusion of AI and high-performance aerial combat platforms. Originally envisioned by Blue Force Technologies as a cost-effective, unmanned aggressor aircraft for training and simulation purposes, the Fury project gained renewed momentum in 2023, when Anduril acquired the design and reimagined it as a fully autonomous, mission-capable system for contested airspace operations.

The Fury's airframe is approximately half the size of an F-16, with a sleek and angular profile tailored for agility and efficiency. It features swept trapezoidal wings, a chin-mounted air intake, and a cruciform tail configuration, emphasizing stability at high altitudes and maneuverability in dynamic aerial environments. Designed for performance just below the speed of sound, the YFQ-44A is expected to reach Mach 0.95 and operate at altitudes up to 50,000 feet, with the structural resilience to withstand up to 9 g-forces — a benchmark typically reserved for manned fighter jets.

Propulsion is provided by a Williams FJ44-4M turbofan engine, generating 4,000 pounds of thrust, offering both efficiency and the power necessary for complex aerial maneuvers. The heart of the Fury's capabilities, however, lies in its brain: Anduril's Lattice AI software suite, which enables autonomous flight control, sensor fusion, threat assessment, and real-time mission decision-making — all without requiring direct human oversight.

As of May 2025, the Fury has entered ground testing, with initial flight trials scheduled for later in the year. While not yet operational, the YFQ-44A represents a new class of AI-enabled aircraft — one designed not only to simulate adversaries for training, but to eventually operate as a loyal wingman, standalone strike platform, or autonomous air superiority asset. In many ways, it signals the future direction of unmanned aerial combat — intelligent, fast, and unburdened by the limitations of a human cockpit.

The **YFQ-42A**, developed by **General Atomics Aeronautical Systems**, represents a significant leap in the evolution of autonomous aerial combat platforms. Rooted in the **XQ-67A demonstrator** from the **Air Force Research Laboratory's Off-Board Sensing Station (OBSS) program**, the YFQ-42A forms part of General Atomics' ambitious **"Gambit" series** — a family of unmanned systems designed to augment and extend the capabilities of manned fifth-generation fighters.

The aircraft features a **sleek, elongated fuselage**, optimized for range and endurance, and incorporates a **dorsal-mounted air intake**, **V-tail configuration**, and **internal weapon bays** to minimize its radar signature and maximize survivability in contested environments. The stealth-optimized airframe is engineered not only for ISR missions but also for direct engagement in combat scenarios.

At its core, the YFQ-42A is a **missile-carrying UAV**, specifically designed to internally house and deploy **two AIM-120 AMRAAM air-to-air missiles.** This internal carriage preserves the aircraft's low-observable profile while offering real kinetic punch in engagements where a human pilot may not be present. As part of its mission profile, the drone is intended to operate seamlessly alongside **F-22 Raptors and F-35 Lightning IIs**, acting as an **off-board sensor node, decoy, or weapons delivery system**, depending on operational needs.

Designed with full integration into the U.S. Air Force's digital kill chain in mind, the YFQ-42A provides forward reconnaissance, threat detection, and strike capabilities — extending the reach of manned aircraft while reducing risk to human pilots.

As of **May 2025**, the YFQ-42A is undergoing **ground testing**, with **flight evaluations anticipated later in the year**. Its development reflects a broader doctrinal shift toward **distributed, networked combat formations**, where unmanned platforms serve not only as support units but as active, intelligent participants in high-intensity conflict.

In **March 2025**, the United States Air Force formally assigned the **"YFQ" designation** to its emerging class of prototype fighter drones, marking a historic milestone in the evolution of autonomous aerial warfare. This designation—traditionally reserved for experimental fighter platforms—signals a clear shift in how unmanned systems are viewed within the broader airpower architecture: no longer as support assets, but as potential peers to manned combat aircraft.

As of **May 2025**, **ground testing** is underway for both the **YFQ-42A** and **YFQ-44A**, with engineers and test pilots focusing on **propulsion performance, avionics systems, autonomy integration**, and **ground control interface reliability**. These early trials are designed not only to validate the technical capabilities of each airframe but to evaluate how they can operate within joint manned-unmanned mission frameworks.

To support future deployment and operational readiness, the Air Force has selected **Beale Air Force Base** in California as the home of the inaugural **Collaborative Combat Aircraft (CCA) Readiness Unit**. This unit will play a pivotal role in shaping training, tactics, maintenance protocols, and live operational drills involving autonomous fighter platforms, preparing for their eventual integration into the U.S. combat fleet.

Strategically, the CCA program reflects a profound transformation in U.S. airpower doctrine. By integrating **autonomous systems** directly into frontline operations, the Air Force aims to enhance **mission flexibility**, extend the survivability of manned assets, and **preserve air superiority** in increasingly contested environments. These drones are expected to execute high-risk or high-tempo missions that would otherwise place human pilots in jeopardy—acting as loyal wingmen, forward scouts, or even autonomous strike platforms depending on mission demands.

As artificial intelligence, sensor fusion, and airframe design continue to evolve, the YFQ series represents not just a technological leap, but a **strategic reimagining of aerial combat itself**—one where humans and machines operate in tandem to dominate the skies of the 21st century.

Chapter 14 - The Tic Tac Dilemma — Drones, Disclosure, and the Limits of Known Technology

In November 2004, 100 miles off the coast of Southern California, a U.S. Navy battle group encountered something no one had seen before — and few were prepared to explain. An object described as a white, featureless, oblong "Tic Tac" was picked up on radar, moving erratically at altitudes and speeds that defied known aerodynamic principles. When two F/A-18 Super Hornet pilots from the USS Nimitz were dispatched to intercept, they visually confirmed an airborne craft with no wings, no exhaust, and flight capabilities far beyond the performance envelope of any known drone or aircraft in existence at the time.

What began as a tightly held military incident would, years later, become a catalyst for a profound public reckoning. When footage of the encounter surfaced in 2017 — released by the U.S. Department of Defense and published by the New York Times — it ignited a wave of speculation and inquiry. Was this a highly classified black project? A foreign surveillance platform? Or something else entirely? Most crucially for national security, why had the world's most advanced naval fleet failed to identify, explain, or respond to an airspace intrusion so dramatic and so close to U.S. shores?

This incident — now widely known as the "Tic Tac encounter" — did not confirm extraterrestrial visitation. But it did something equally important: it exposed the limits of our current airspace monitoring systems, especially when it comes to distinguishing between advanced drones, unmanned surveillance vehicles, and true unknowns. It became a watershed moment not just for UFO research, but for UAV policy and military accountability.

For decades, unmanned aerial systems had operated in the margins — used by militaries, intelligence services, and in recent years, commercial industries. But the Tic Tac incident underscored a deeper problem: what happens when technology outpaces detection? When objects fly without transponders, markings, or exhaust signatures — and when they maneuver in ways that defy established norms — even the most sophisticated radar and tracking systems can fail to categorize them. In this sense, the Tic Tac encounter was less about what the object was, and more about what it revealed — that the U.S. military was not adequately prepared to classify or respond to all aerial incursions, especially those without human pilots onboard.

Since that day, a quiet revolution has taken place. Government agencies have launched new task forces, Congress has held public hearings, and the defense industry has been pressured to innovate around detection, tracking, and classification — all sparked, in part, by a six-meter-long white object dancing over the Pacific Ocean.

This chapter explores the intersection of unmanned systems, unidentified phenomena, and the policy and technology gaps exposed by the 2004 incident. In doing so, it reveals not just a single mystery — but a broader demand for transparency, accountability, and clarity in the age of drones and data-driven defense.

On November 14, 2004, the U.S. Navy's Carrier Strike Group 11 was conducting routine training exercises off the coast of Southern California when a series of anomalous radar contacts disrupted the operation. For several days prior, the USS Princeton, a Ticonderoga-class guided missile cruiser, had been tracking unknown aerial targets that appeared suddenly at 80,000 feet, descended rapidly to sea level in under a second, hovered for short periods, and then vanished from radar. The radar operators initially believed the system was malfunctioning — until visual confirmation came from the air.

Two F/A-18F Super Hornet fighter jets launched from the USS Nimitz were dispatched to investigate. Commanding the flight was Commander David Fravor, a seasoned naval aviator with nearly two decades of experience. Accompanying him was Lieutenant Commander Jim Slaight, the weapons systems officer in the back seat. Upon arrival at the designated intercept coordinates, Fravor noticed that the sea surface below was churning, as though something large was submerged. Hovering just above this disturbance was a smooth, white, oblong craft — approximately 40 feet long, shaped like a "Tic Tac" breath mint. It had no wings, no rotors, no visible means of propulsion, and it moved in ways no aircraft they had ever encountered could.

As Fravor attempted to engage, the object reacted with uncanny awareness and agility — mirroring his movements, then accelerating instantly out of sight. "It just zipped off like nothing I've ever seen," he later told interviewers. The speed and maneuverability exceeded anything available to U.S. or foreign militaries at the time.

Later that same day, a subsequent flight captured infrared video of the object using the **Advanced Targeting Forward-Looking Infrared (ATFLIR)** system. The resulting footage, which depicted a **Tic Tac-shaped craft** executing rapid maneuvers and exhibiting flight characteristics inconsistent with known aerodynamic principles, would remain **classified and undisclosed to the public for over a decade**. Internally, however, the incident was taken seriously—**documented, analyzed, and preserved within military channels** as an anomalous engagement of unknown origin.

What made the encounter so remarkable was the **multi-sensor confirmation** it received. The **radar array aboard the USS *Princeton*** tracked the object at distances and speeds that defied known propulsion systems. **F/A-18 pilots** visually observed the craft in clear airspace, reporting its smooth, white exterior and lack of visible control surfaces or exhaust. The ATFLIR pod provided **thermal imagery**, confirming the object's heat signature and erratic movement. Meanwhile, **radio communications** and tactical data links corroborated the pilots' experience in real time, offering further validation of the event's authenticity.

Despite this convergence of evidence—from radar to optical to infrared to voice logs—there was **no aircraft in the U.S. inventory** capable of matching the object's performance profile. Nor was there any **credible indication of foreign adversaries** possessing such advanced aerospace technologies. At least none that had been **publicly acknowledged**. The event, initially buried in official channels, would remain a **classified enigma** until it was thrust into the public spotlight in **2017**, becoming one of the most scrutinized and compelling incidents in the modern study of unidentified aerial phenomena.

What made the incident so compelling was not just the behavior of the object, but the credibility of the observers and the integrity of the sensor data. The event had all the elements of a tightly controlled military encounter: secure systems, chain of command, and corroborating testimony from trained professionals. And yet, no one could explain what it was.

For over a decade, the encounter was whispered about in defense circles and aviation forums but remained largely under wraps. It wasn't until a group of former Pentagon officials and intelligence insiders — including Luis Elizondo, who had run the Department of Defense's Advanced Aerospace Threat Identification Program (AATIP) — began pushing for disclosure that the Tic Tac incident reached the public. In 2017, the New York Times broke the story, accompanied by officially released Navy footage, confirming what many had long suspected: something unclassified, unexplained, and potentially disruptive had occurred in U.S. airspace.

The Tic Tac incident wasn't merely a curiosity. It was a paradigm shift. It forced military and intelligence communities to acknowledge what had long been avoided — that not all aerial threats fit within known categories, and not all unknowns can be dismissed as fiction. In the wake of that encounter, the conversation about drones, UAPs, and the surveillance of American airspace would never be the same.

When the details of the 2004 Tic Tac encounter were made public, a key question immediately arose: could the object have been a highly advanced unmanned aerial vehicle (UAV)? Perhaps a black project operated by a U.S. agency, a clandestine test by a foreign power, or a next-generation drone so advanced that it defied public understanding? To answer that, it's necessary to compare the reported capabilities of the Tic Tac craft with the known state of drone technology at the time.

In 2004, the U.S. military's drone capabilities were already revolutionizing modern warfare, particularly in the Middle East. The General Atomics MQ-1 Predator, armed with AGM-114 Hellfire missiles, and the Northrop Grumman RQ-4 Global Hawk, optimized for high-altitude surveillance, had become central to U.S. intelligence and strike operations. These UAVs were capable of loitering over targets for extended periods, transmitting real-time video, and executing precision strikes—all while keeping pilots out of harm's way.

Yet despite their effectiveness, these systems were not without limitations. Their performance was governed by the physics and engineering of conventional aircraft. None of the operational drones in 2004 could reach supersonic speeds, let alone perform the kind of instantaneous acceleration or abrupt vertical ascents from sea level to the upper atmosphere that would later be associated with the 2004 Nimitz encounter. Their maneuverability, too, was designed around steady, fuel-efficient flight paths; abrupt directional changes, high-G turns, or hover-like behavior were simply beyond their aerodynamic envelopes.

Stealth was another evolving area. While low-observable technologies were beginning to shape the design of next-generation UAVs, the drones of the early 2000s remained susceptible to radar detection—especially by sophisticated phased-array systems like those aboard the USS *Princeton*. And all known UAVs at the time, whether powered by turboprops, turbojets, or piston engines, produced distinct heat and sound signatures, easily picked up by thermal sensors or audio surveillance.

By contrast, the object observed by Navy personnel during the USS *Nimitz* incident—often described as a "Tic Tac"—exhibited flight characteristics that defied not just drone capabilities but the known laws of aerodynamics and propulsion. Witnesses and sensor systems recorded a craft with no visible control surfaces or propulsion, capable of hovering silently and then accelerating to hypersonic speeds in an instant. The craft transitioned seamlessly between altitudes, including rapid descents from 80,000 feet to sea level, without creating sonic booms, heat signatures, or turbulence—phenomena that would accompany any known man-made vehicle under such conditions.

Given these capabilities, speculation has arisen that the object may have belonged to a deeply classified U.S. "black" program—perhaps an advanced prototype UAV undergoing operational testing in live conditions. The American military-industrial complex has a well-documented history of concealing breakthrough aerospace projects, with programs like the U-2, A-12, and SR-71 Blackbird developed in total secrecy at facilities like Area 51. It is plausible, some argue, that the Tic Tac was a next-generation aerospace vehicle so advanced that even the Navy pilots and radar operators present during the incident were not cleared to know of its existence.

However, multiple factors cast serious doubt on the black-project hypothesis. Most notably, the incident involved multiple military assets operating in a secure training area, and no warnings or restrictions had been issued regarding ongoing flight tests—standard procedure even for classified programs. Furthermore, the object's performance—specifically the lack of any observable propulsion, inertia-defying maneuvers, and its apparent ability to anticipate and respond to fighter movements—goes well beyond the known capabilities of even the most speculative U.S. aerospace concepts at the time.

There is also a broader question of strategic logic: it would be highly unorthodox for a secret U.S. program to risk exposure by testing such a system in full view of operational carrier strike groups, with pilots, radar technicians, and intelligence personnel all unaware of its presence. If it were a black project, it was one strangely deployed, seemingly without regard for compartmentalization or security protocol.

Thus, while the black-program theory remains one of the few conventional explanations still considered plausible by some analysts, the preponderance of evidence suggests that the object seen during the 2004 Nimitz encounter represented flight characteristics that—by the standards of known aerospace technology in 2004—were not just classified, but fundamentally unprecedented.

First, conducting such a test **in the middle of a live naval exercise**—without notifying the commanders of the **USS Nimitz Carrier Strike Group**—would represent a severe breach of military protocol. It would endanger both personnel and equipment and risk a **potentially catastrophic escalation**, should the unknown craft be engaged as a perceived threat. Second, the **multi-domain exposure** of the object—detected simultaneously by **radar, infrared sensors, visual observation, and tactical communications systems**—is inconsistent with the standard practice of concealing advanced technologies from broad detection. Black programs are typically designed to operate in **controlled, low-observability environments**, not ones where the risk of exposure is amplified.

Finally, **multiple senior Navy officers** involved in the incident, including commanding officers and radar specialists, have gone on record stating that **had this been a known U.S. platform**, they would have been briefed accordingly. **They were not**. Their consensus has remained firm: **this was not a known asset**.

Could the Tic Tac have belonged to a **foreign adversary**, such as **Russia or China**? While theoretically possible, the technical evidence and strategic context strongly argue against it.

In **2004**, neither nation had demonstrated capabilities approaching those observed in the incident. Both were in the early stages of developing large surveillance drones and **loitering munitions** — platforms that prioritized endurance or target tracking, not extreme maneuverability or stealth. No known Russian or Chinese UAV from that era employed **non-lift-based propulsion**, nor could they achieve the **rapid acceleration, high-altitude maneuvering, and evasive agility** reported by Navy personnel.

Moreover, the idea that a foreign power would **deliberately insert a cutting-edge prototype** into restricted U.S. military training waters — during a major naval exercise — strains credulity. Such an act would risk **high-profile detection**, potential recovery of the craft, and serious international consequences. It would be, in essence, a **strategically reckless move**, offering little intelligence value while inviting a diplomatic crisis.

Finally, even two decades later, **no nation has unveiled any technology** that matches the performance attributed to the Tic Tac. In an era where hypersonic weapons, stealth drones, and unmanned combat systems are proudly showcased for deterrence and prestige, the continued **absence of any parallel platform** is conspicuous. If such a capability existed, its strategic utility would be immense—and yet, **there has been silence**.

Internal Pentagon briefings and leaked reports from former officials suggest that even classified threat assessment units were unable to attribute the Tic Tac object to any nation or known program. Its observed characteristics remained "unidentified" in both origin and capability, with some analysis suggesting it demonstrated non-inertial movement inconsistent with any terrestrial propulsion system.

Thus, while many initially tried to fit the Tic Tac within the framework of UAV evolution, its capabilities so vastly exceeded the known limits of drone and aerospace engineering at the time that it spurred a more fundamental question: What, exactly, was it — and who controls it?

For over a decade following the 2004 Tic Tac encounter, the incident remained largely confined to classified channels within the U.S. military and intelligence community. It was discussed quietly among pilots, radar technicians, and intelligence officers — respected as a real, well-documented event but one without a public forum. That changed dramatically in December 2017, when The New York Times published an explosive front-page article titled "Glowing Auras and 'Black Money': The Pentagon's Mysterious U.F.O. Program."

The article revealed the existence of the Advanced Aerospace Threat Identification Program (AATIP) — a shadowy Pentagon unit tasked with investigating aerial anomalies. Directed by intelligence officer Luis Elizondo, AATIP had operated under the Defense Intelligence Agency (DIA) with funding secured by former Senate Majority Leader Harry Reid. The Tic Tac incident was featured prominently in the piece, alongside officially released Navy footage — the now-famous FLIR video — showing a strange, white object darting across the screen at high speed, tracked by targeting sensors but defying easy explanation.

For the first time, the subject of **unidentified aerial phenomena (UAPs)** entered the **mainstream discourse** with credibility and urgency. A convergence of factors—**military-verified footage, veteran naval witnesses, and coverage by reputable media outlets**—brought the 2004 **Tic Tac incident** from classified obscurity into public awareness. The **Department of Defense** officially confirmed that the videos released to the press were **authentic**, captured by **U.S. Navy pilots** using onboard infrared targeting systems. Yet, the Pentagon offered **no formal explanation** as to the identity or origin of the object.

Its statements were **deliberately measured**, acknowledging that the object was **real**, **unidentified**, and had **violated military-controlled airspace**. The absence of a clear attribution—coupled with official confirmation—only intensified interest. For many, this marked a **turning point**: the U.S. government was no longer dismissing or denying such encounters, but was instead **admitting the existence of aerial anomalies** that defied current explanation.

The result was a **profound cultural and institutional shift**. What had long been relegated to **fringe theories and science fiction** was now **a topic**

of legitimate national security concern. The language itself evolved: the term **"Unidentified Flying Object" (UFO)** gave way to **"Unidentified Aerial Phenomenon" (UAP)** — a rebranding intended to distance serious inquiry from pop culture associations and reposition the issue within **aerospace and defense frameworks.**

The public narrative was further reshaped by **former military personnel** who stepped forward. **Commander David Fravor**, the lead pilot who visually encountered the Tic Tac off the coast of Southern California, became a prominent voice. Alongside other Navy aviators and radar operators, Fravor appeared in televised interviews, participated in investigative documentaries, and testified in **classified congressional briefings.** Their accounts—delivered with the credibility of combat-experienced officers—lent weight to what had once been dismissed as implausible.

These developments reframed the UAP conversation from **speculation to structured investigation.** The focus shifted toward **sensor-verified military encounters**, **corroborating radar data**, and **official government acknowledgment**, even if that acknowledgment came without answers. What remained unsolved now also remained undisputed: something unknown had entered restricted U.S. airspace, had been observed by trained pilots, recorded by advanced targeting systems, and **recognized—though not explained—by the Pentagon itself.**

The public pressure triggered by the Tic Tac disclosures compelled Congress to act. In 2020, the Senate Intelligence Committee — then chaired by Senator Marco Rubio — requested a comprehensive report on UAPs from the Director of National Intelligence (DNI) and the Pentagon. The resulting June 2021 report acknowledged 144 credible military UAP incidents since 2004, with only one definitively explained. While the report refrained from drawing conclusions about origin, it highlighted the potential for technological surprise from foreign adversaries, or worse — unknown sources.

DRONE INTRUSIONS
OVER MILITARY
INSTALLATIONS

DRONE INCIDENTS AT U.S. MILITARY SITES

Recent Unidentified Drone Detections at Sensitive U.S. Military Installations

Increase in Reported Incursions

- Rise in reported drone incursions overe restricted facilities such as nuclear laboratories. missile storage complexes
- Many incidents often ail concerns among security officicars
- Ongoing investigations by U.S government ongo into origins and purposes of these drones

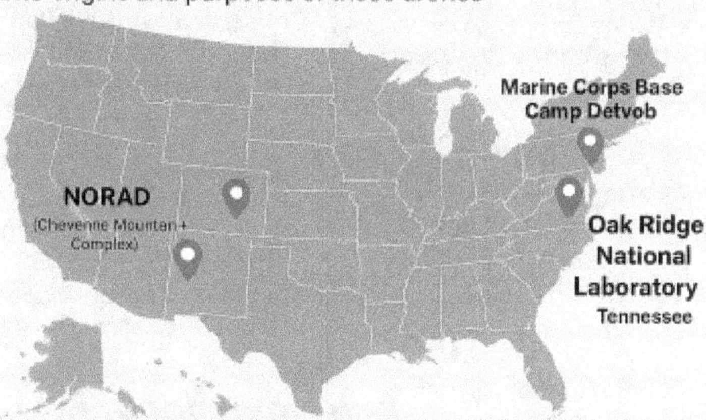

Potential Threats and Risks

- Espionage: activities aimed atcollecting sensitive information
- Sabotage: potential attakack on critical infrastructure
- Enhancement alversary capabilities:"drone swarms"
- Need: to further strengthen security measures at itly sites

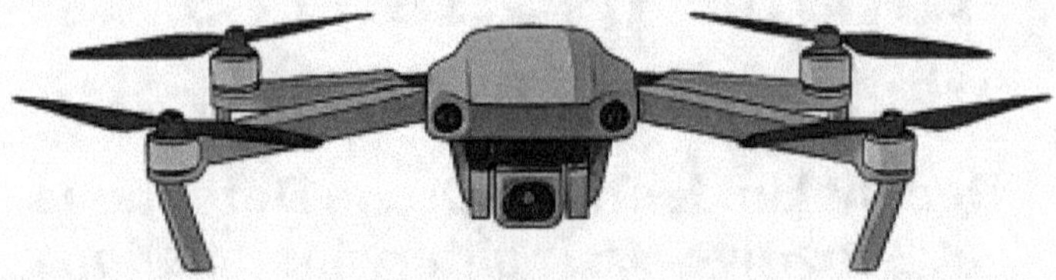

ESCALATING THREATS–

DRONE OVERFLIGHTS AT U.S. MILITARY FACILITIES

APR 2019
Numerous drone flights over Marine Corps Base over-flights via The War Zone

MAY 2019
Six unauthorized swarms observed over USS Paul Hamiiton and USStunker Hill off the California coast

SEP 2019
As many as six drones loi-tering and later disappeaar near the the Pentagn

JUL 2021
Reports of drone swarms observed over the USS Paul Hamilton an USS Bunker Hill off the California coast

JUL 2021
Reports of drone swarms observed over over USS Paul Hamilton and USS Bunker coast

SEP 2023
Eight drones flew over Palo Verde Nuclear Generating Stationin Arizona and unknown drone events investigated by U.S, goverriment as UAPincidents

MAY 2023
Six unauthorized small drones flew over Joint Base Lewis-McChord in Washington

APR 2024
Multiple drones of unknown origin buzzed around Naval Base Coronado in California

CHAPTER 12

Drone Incidents Over Military Installations

In early 2024. U.S military installations reported an increase in unauthorized drones seen near facilities. These encounters have raised alarms about security risks and prompted feder
al investigations.

Pentagon Warning

"We have seen an increase in drones around our military installations... which poses a potential threat to our service members."

Sabrina Singh, Pentagon spokesperson **February 7, 2024**

Feb 4

Hill Air Force Base, Utah

Base security and local authorities pursued five drones that hovered near the installation before departing

Feb 5

Davis-Monthan Air Force Base, Arizona

Two drones flew into airspace and near an aircraft hangar

Feb 7

Joint Base Charleston, South Carolina

A drone described as larger than others previously observed vai sight near the base

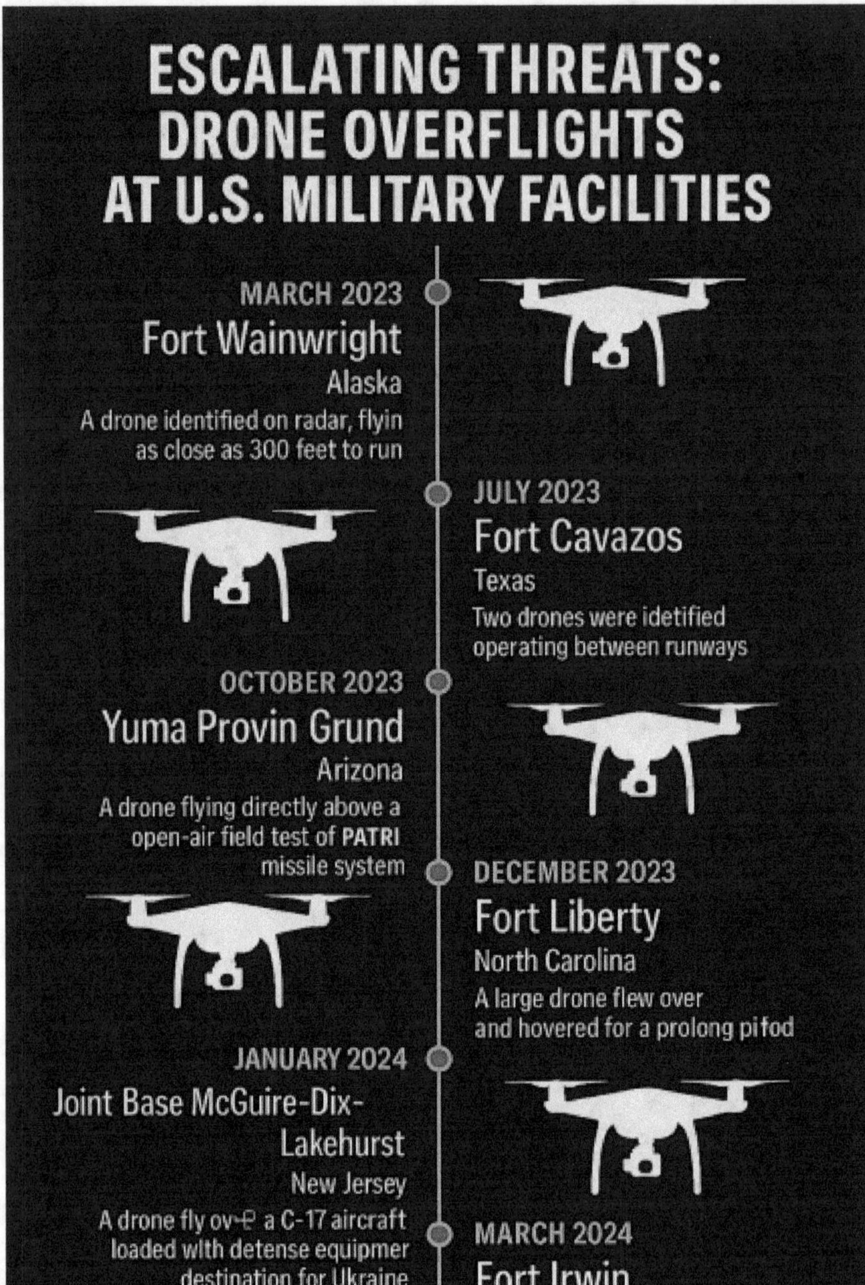
ESCALATING THREATS:
DRONE OVERFLIGHTS
AT U.S. MILITARY FACILITIES
MARCH 2023
Fort Wainwright
Alaska
A drone identified on radar, flyin
as close as 300 feet to run
JULY 2023
Fort Cavazos
Texas
Two drones were idetified
operating between runways
OCTOBER 2023
Yuma Provin Grund
Arizona
A drone flying directly above a
open-air field test of PATRI
missile system
DECEMBER 2023
Fort Liberty
North Carolina
A large drone flew over
and hovered for a prolong pifod
JANUARY 2024
Joint Base McGuire-Dix-
Lakehurst
New Jersey
A drone fly ov-e a C-17 aircraft
loaded with detense equipmer
destination for Ukraine
MARCH 2024
Fort Irwin

In 2022, the Department of Defense took an unprecedented step by establishing the All-Domain Anomaly Resolution Office (AARO), a permanent federal entity tasked with investigating, cataloging, and assessing anomalous phenomena observed in the air, at sea, and in space. For the first time in U.S. history, the government formally committed to the study of objects that defied classification—craft that were neither conventional aircraft, drones, balloons, nor any other known human technology. This institutional shift reflected a growing realization within the defense and intelligence communities that some encounters reported by trained military personnel could not be easily dismissed.

Among these, the 2004 Nimitz incident—centered around the now-infamous "Tic Tac" object—stood out not only for its richness in eyewitness accounts and sensor data, but for the profound ripple effects it generated. What distinguished the Tic Tac was not simply its flight profile or the credibility of those who observed it, but the way it forced open the door to official U.S. government acknowledgment of unidentified aerial phenomena (UAPs). It marked a turning point—a moment when speculation gave way to documentation, and when reports once buried in classified files emerged into the daylight of congressional hearings and public scrutiny.

The encounter lent legitimacy to what had long been stigmatized. Pilots, radar operators, and intelligence officers—individuals trained to interpret anomalies—were suddenly able to speak openly, without fear of ridicule or career consequences. Scientific institutions began to explore the subject with renewed seriousness, and defense policymakers were compelled to reassess how threats to sovereign airspace were defined, identified, and responded to.

In this light, the Tic Tac incident evolved from a misunderstood radar anomaly to something far more consequential: a catalyst for a new era of disclosure, inquiry, and reform. It reframed the conversation from fringe theory to institutional challenge—no longer a question of belief, but one of policy, science, and national preparedness. It remains the case that no one knows exactly what the object was. But what is now undeniable is what it changed.

The public release of the Tic Tac incident, along with two other Navy videos (GIMBAL and GOFAST), prompted a fundamental change in how the U.S. government approached the question of unidentified aerial phenomena (UAPs). Previously, military encounters with unidentified objects were often dismissed, buried in paperwork, or confined to classified reporting channels. But the credibility of the Tic Tac witnesses and the multiple systems that tracked the object — visual, infrared, radar, and telemetry — forced defense officials and lawmakers alike to acknowledge that something unprecedented had occurred.

After the 2017 revelations, former AATIP director Luis Elizondo publicly criticized the Pentagon for its lack of transparency and seriousness regarding the UAP issue. He resigned in protest, stating that the government was failing to adequately investigate what he viewed as potential airspace incursions by unknown vehicles exhibiting advanced flight capabilities.

In response to mounting public interest and pressure from former insiders, the Pentagon established the UAP Task Force in August 2020 under the Office of Naval Intelligence. Its mission: to "detect, analyze, and catalog UAPs that could potentially pose a threat to U.S. national security." Unlike AATIP, which operated in relative secrecy, the UAP Task Force was authorized with the backing of Congress, reflecting a shift from passive awareness to active oversight.

In June 2021, the veil of official silence surrounding unidentified aerial phenomena began to lift when the Office of the Director of National Intelligence (ODNI), acting under direction from the Senate Intelligence Committee, delivered a long-awaited unclassified report to Congress. Though measured in tone and deliberately cautious in its conclusions, the report marked a watershed moment in the modern study of UAPs. Out of 144 military encounters reviewed between 2004 and 2021, only one was definitively explained—a startling acknowledgment that the vast majority remained unexplained, even after investigation by some of the most advanced intelligence-gathering apparatuses in the world.

More striking than the numbers were the flight characteristics attributed to these objects. Several were reported to exhibit performance beyond known aerospace capabilities—demonstrating sudden acceleration, the ability to hover midair without visible propulsion, and the astonishing feat of transmedium travel, seamlessly moving between air and water without turbulence or structural disruption. These anomalies, recorded by trained observers and sophisticated sensors, could not be easily dismissed as conventional aircraft, drones, or atmospheric illusions.

While the report stopped short of suggesting extraterrestrial origins, it also notably refrained from attributing the sightings to classified U.S. programs—an omission that carried its own weight. The document warned that these objects could reflect technological advancements by foreign adversaries or, more provocatively, from "other sources" not yet understood. That ambiguity prompted alarm in national security circles and renewed calls for systematic analysis, risk assessment, and oversight.

The response in Washington was swift. In July 2022, Congress convened its first public hearings on UAPs in more than half a century. The bipartisan tone was clear: this was not a discussion about flying saucers, but about sovereignty, security, and the unexplained presence of advanced aerial platforms operating with impunity near sensitive military installations. Lawmakers pressed Pentagon officials on the absence of a coherent investigative framework and demanded greater transparency, not just for military accountability but for public trust.

Later that year, the Department of Defense took a decisive step to meet that demand. As part of the 2022 National Defense Authorization Act, it established the All-Domain Anomaly Resolution Office (AARO), a permanent successor to the UAP Task Force. Unlike its predecessor, which had a narrow and temporary scope, AARO was granted a far-reaching mandate: to identify, analyze, and assess anomalies not just in the air, but also in maritime, spaceborne, and subsurface domains. This reflected a growing understanding that the phenomena under investigation might not be limited to conventional airspace—or to conventional explanations.

With AARO, the United States formally institutionalized the search for answers to one of the most enduring and elusive challenges in modern defense. What was once relegated to rumor, ridicule, or compartmented briefings had now entered the mainstream of national policy. The age of institutional engagement with UAPs had begun—not with sensationalism, but with an insistence on data, rigor, and accountability.

The establishment of the All-Domain Anomaly Resolution Office marked more than just bureaucratic evolution—it represented a foundational shift in how the U.S. government approaches the unknown. AARO's mission was deliberately broad and ambitious. It was charged with bridging the fragmented landscape of anomalous aerial analysis, uniting intelligence agencies, military branches, and civilian institutions under a single mandate. For the first time, the Department of Defense committed to a standardized system for documenting, evaluating, and responding to unidentified phenomena—ensuring that credible sightings were neither dismissed out of hand nor lost in the shuffle of conflicting reporting protocols.

One of AARO's earliest tasks was to bring uniformity to what had previously been a patchwork of anecdotal reports, isolated radar contacts, and uncorrelated data streams. It developed structured reporting procedures, enabling pilots, radar operators, and satellite technicians to submit observations into a centralized database for review. In parallel, the office worked to triage these reports—sorting natural or technological misidentifications from those requiring deeper scrutiny. Of particular interest were encounters involving advanced sensor corroboration, in which visual observation was matched by radar, infrared, or electro-optical confirmation.

AARO's creation acknowledged that UAPs were not merely the domain of speculative fiction or fringe science—they were increasingly a national security concern. The rise of small, stealthy drones and autonomous flight systems had blurred the line between ordinary surveillance assets and extraordinary aerial behavior. Military planners recognized that a failure to distinguish between foreign UAV incursions and true anomalies could leave critical vulnerabilities unaddressed.

Thus, AARO's work unfolded against a backdrop of urgent questions. Could some sightings be misclassified drones—small, maneuverable, and low-observable? Were adversaries deploying surveillance systems that eluded traditional detection methods? And what, if anything, remained that could not be reconciled with known physics or available aerospace technology?

What set AARO apart from previous efforts was its interdisciplinary approach. It did not rely solely on intelligence analysts or defense insiders, but opened its doors to scientists, aerospace engineers, propulsion experts, and artificial intelligence specialists. These professionals were brought in to analyze telemetry, motion profiles, and sensor signatures, applying mathematical modeling and high-fidelity simulations to determine whether observed behavior could be replicated—or explained—within the limits of current engineering.

This pivot toward scientific inquiry was a direct response to decades of cultural and institutional reluctance to treat UAPs with seriousness. Events like the Tic Tac encounter—recorded by trained pilots and confirmed by infrared targeting systems and radar arrays—had forced a reckoning. The old paradigm of ridicule and dismissal could no longer hold in the face of empirical anomalies backed by sensor data.

By inviting rigorous, peer-informed scrutiny of UAP incidents, AARO redefined the conversation. Its formation marked not only a change in policy, but a shift in cultural posture: from secrecy to transparency, from anecdote to analysis, from silence to study. The skies, once regarded as domains of certainty, had revealed unknowns that demanded a new framework—and AARO was the response. In an age increasingly shaped by drones, artificial intelligence, and near-peer technological competition, the agency stands at the frontier where defense, science, and mystery converge.

The Tic Tac incident did more than trigger official investigations — it ignited a widespread call for transparency in both the public and private sectors. For decades, the topic of unidentified aerial phenomena had been dismissed as fringe or speculative. But the credibility of military witnesses, sensor data, and government confirmations in the aftermath of the Tic Tac disclosures forced a reckoning: if these objects exist and remain unexplained, the public has a right to know how they are being studied — and how airspace sovereignty is being defended.

U.S. lawmakers across the political spectrum began demanding regular briefings and unclassified reporting on UAP encounters. Members of the Senate Intelligence and Armed Services Committees, including Senator Marco Rubio, Senator Kirsten Gillibrand, and Representative Tim Burchett, emerged as vocal advocates for government accountability.

The mounting concern within government circles regarding unidentified aerial phenomena came to focus around three pivotal issues: national security, technological surprise, and institutional credibility. At the core was the realization that unknown objects were breaching restricted military airspace without interception or identification—an intolerable scenario for any defense infrastructure. These incursions suggested potential vulnerabilities not only in detection systems, but in the broader protocols that governed the nation's ability to respond to aerial threats.

Compounding the unease was the specter of technological surprise. If the objects observed during military encounters—such as the now-famous Tic Tac incident—were not of U.S. origin, then the possibility loomed that near-peer adversaries like China or Russia had quietly achieved revolutionary advancements in propulsion or drone technology. Such a leap, if true, would represent a profound failure of American intelligence to forecast and counter a strategic threat.

Beyond defense implications, lawmakers began to question whether the government had withheld knowledge or capabilities related to UAPs from both the public and its elected representatives. The issue of transparency—long the Achilles' heel of UAP discourse—reasserted itself with urgency. If data existed that could clarify or contextualize these sightings, then democratic accountability demanded its disclosure.

In response to these concerns, Congress passed legislation requiring annual unclassified reports on UAPs from the Department of Defense, along with regular briefings to lawmakers. More significantly, the law mandated standardized reporting protocols across all branches of the U.S. military—something that had never been formalized despite decades of sporadic encounters. For the first time, pilots and radar operators were encouraged, even required, to report anomalies without fear of stigma or reprisal.

But even as data collection improved, one challenge grew more acute: the problem of attribution. It was no longer enough to simply acknowledge that anomalous objects had been detected. Analysts needed to determine what these objects were—and just as crucially, what they were not.

This task, once the domain of fringe theorists and defense contractors operating in the shadows, became a matter of national urgency. In a modern airspace increasingly filled with drones, balloons, commercial UAVs, and covert reconnaissance assets, the process of identification has grown exponentially more complex. A given sighting might be a military drone on a classified mission, a commercial quadcopter straying into restricted airspace, a foreign surveillance platform operating without transponder identification, or something far less conventional.

Attribution, then, is not merely a technical challenge—it is a strategic necessity. Without knowing the origin and purpose of what's in the sky, the risk of miscalculation increases, as does the potential for missed threats. Misidentifying a cutting-edge foreign UAV as a benign commercial drone—or vice versa—could have dire consequences. Likewise, mistaking genuinely unclassified phenomena for hostile platforms could trigger unnecessary alarm or inappropriate response.

As sightings multiply and detection systems become more sophisticated, the imperative to classify aerial anomalies with precision only grows. It is within this evolving context that agencies like the All-Domain Anomaly Resolution Office (AARO) now operate—tasked not just with cataloging unknowns, but with untangling them from the known. In an era where both advanced adversaries and unexplained capabilities share the same sky, the future of aerospace security will hinge not just on what can be seen—but on what can be clearly understood.

To address this growing ambiguity, a chorus of **lawmakers, defense officials, and researchers** has called for the development of a **national UAV identification protocol**. This system would aim to tag and track **authorized drones** via **digital identifiers**, much like aircraft transponders, allowing airspace managers to quickly validate friendly or commercial operations. Simultaneously, the system would be designed to **flag aerial anomalies**—those that fail to conform to recognized flight paths, known vehicle signatures, or expected telemetry.

Critically, such data would be **fed into centralized military and civilian databases** in real time, facilitating cross-agency coordination between the **Department of Defense**, **FAA**, **NORAD**, and domestic law enforcement. This would allow for a faster, more accurate response to aerial intrusions and provide a clearer framework for assessing whether an unknown object poses a threat, represents a technological breakthrough, or simply reflects an identification failure.

This push for **classification clarity** reflects a new strategic reality: **in the drone age, uncertainty is a threat vector of its own**. Not knowing what an object is—or who controls it—can be as dangerous as the capabilities it may possess. As the skies grow more crowded with autonomous systems, ensuring that every airborne entity is properly tagged, tracked, and attributed may become the cornerstone of airspace security in the 21st century.

The Tic Tac incident also galvanized academia, think tanks, and scientific institutions. Once hesitant to engage with UAPs, researchers began calling for open data access and independent verification of government claims. Organizations such as the Scientific Coalition for UAP Studies (SCU) and individuals affiliated with institutions like Harvard, Stanford, and MIT began analyzing declassified videos and sensor metadata to assess physical parameters — acceleration rates, thermal signatures, and electromagnetic behavior.

Their push echoed a growing sentiment: if government agencies are collecting data on objects that defy explanation, those data sets should be accessible to independent scientific review.

In parallel, a number of former officials — including AATIP personnel, intelligence officers, and defense contractors — began speaking publicly or testifying behind closed doors about alleged efforts to conceal UAP programs, suppress witness reports, or withhold materials.

This crescendo of insider disclosures has contributed to new legislation protecting whistleblowers who come forward about UAP-related technologies, reverse engineering efforts, or crash retrievals. While some claims remain unverified, the volume and consistency of insider accounts suggest a deeper history of secrecy that is now unraveling under public pressure.

The Tic Tac incident, once dismissed as an isolated anomaly, now stands as a turning point in the demand for airspace transparency. It reframed the conversation from "Is it extraterrestrial?" to "Why aren't we being told what's operating in our skies — and who is responsible for it?"

In the aftermath of the Tic Tac revelations, public and institutional interest in UAPs surged — but so did confusion. The rapid advancement and proliferation of drone technology in both military and civilian sectors complicated the issue in unexpected ways. In many cases, drones and UAVs became indistinguishable from UAPs, especially when operated without lights, identification signals, or known flight patterns.

By the 2020s, the skies above cities, towns, and critical infrastructure were no longer the exclusive domain of licensed aircraft and military systems. Consumer drones, once the domain of hobbyists and tech enthusiasts, had proliferated at a staggering pace—flown by filmmakers, activists, journalists, delivery services, and increasingly, by criminals. On any given day, thousands of quadcopters buzzed through the air, many equipped with GPS, high-resolution cameras, autonomous navigation systems, and flight ranges that rivaled small aircraft. This democratization of aerial technology brought unprecedented access to the sky—but it also brought chaos.

With so many aerial platforms in operation, the airspace grew not only busier but more opaque. Reports of unidentified aerial objects surged—not necessarily because something otherworldly was occurring, but because the sheer density of low-flying, low-visibility drones made attribution nearly impossible. Law enforcement and emergency responders frequently encountered drones near active crime scenes, wildfires, or major traffic accidents. In most cases, no operator could be identified. Were they journalists? Bystanders? Criminal scouts? The answers were rarely clear, and the regulatory response often lagged behind the pace of adoption.

Commercial airliners, particularly those approaching major airports, began filing increasing numbers of near-miss reports. Pilots described encountering small, fast-moving objects at altitudes where drones were legally prohibited. These sightings often turned out to be consumer quadcopters flown illegally and dangerously near controlled airspace. But without transponders or identifying markers, there was no way to know for certain in the moment.

More troubling were the incursions over military facilities, especially those housing nuclear weapons or other strategic assets. Base security teams reported swarms of small UAVs conducting pattern flights, hovering near launch silos, or flying low over perimeter fences—incidents that triggered high-level investigations and heightened alert postures. In these cases, the drones were real, the intent unknown, and the implications deeply unsettling. Were these acts of espionage? Testing of response times? Or simply rogue hobbyists with no sense of boundaries? The ambiguity itself was part of the threat.

As the skies filled with drones, the distinction between UAPs and known UAVs began to erode. Not every anomalous sighting was extraordinary—but not every drone was harmless. The presence of cheap, maneuverable, and often anonymous flying machines blurred the once-clear categories that had guided aerial defense and law enforcement. What emerged was a new paradigm: one in which an unclassified object in the sky could represent anything from a recreational device to a hostile surveillance platform, or something beyond current understanding altogether.

This convergence of technology, behavior, and uncertainty has complicated not only security protocols, but the very concept of airspace control. The UAP phenomenon—long associated with the extraordinary—has been reframed in the 21st century as much by the proliferation of the ordinary as by the presence of the unexplained.

The rise of increasingly sophisticated drone technologies—both domestic and foreign—has profoundly complicated the quest to understand unidentified aerial phenomena. In particular, the United States' long history of classified aircraft development under "black" programs offers a compelling yet controversial explanation for some UAP sightings. For decades, the U.S. military has fielded capabilities far ahead of what is publicly acknowledged, often hidden behind layers of special access programs and need-to-know security. From the stealth fighters of the 1980s to the still-rumored hypersonic testbeds of today, these platforms push the boundaries of known aerospace performance, and in doing so, blur the lines between the possible and the extraordinary.

Some analysts argue that craft described in certain UAP encounters—those exhibiting rapid acceleration, radar evasion, or silent maneuvering—might not be otherworldly at all. Instead, they suggest these sightings may represent next-generation UAVs still buried within classified programs: hypersonic reconnaissance drones capable of Mach 5-plus flight, solar-powered UAVs that linger for weeks in the stratosphere, or stealthy loitering munitions designed to cruise undetected over denied territory. Such systems, designed for high-endurance surveillance, electronic warfare, or kinetic strikes, would naturally exhibit performance characteristics that defy conventional expectations and evade routine identification.

At the same time, foreign drone programs have evolved with alarming speed. China, Russia, Iran, and Turkey have all demonstrated novel UAV technologies ranging from AI-directed surveillance platforms to drone swarms capable of overwhelming advanced air defenses. In this new era of strategic competition, the use of drones not merely as weapons or scouts, but as tools of deception, is no longer hypothetical. Adversaries may deliberately simulate UAP-like behavior—abrupt changes in velocity, sharp angular turns, or momentary radar invisibility—not to mimic aliens, but to test American defenses, gauge radar sensitivity, and gather data on response patterns. These operations would fall squarely within the playbook of electronic warfare and strategic misdirection.

The consequences of such misidentification are serious. If state actors can engineer drones to appear as unidentifiable anomalies—either to cloak their true intent or to sow confusion—they gain a powerful asymmetric tool. It allows them to operate within sensitive airspace, challenge sovereignty, and undermine deterrence, all while denying responsibility. In this context, the UAP issue becomes less about the unexplained and more about the unacknowledged—either because the platform is classified, foreign, or masquerading as something it's not.

This reality reframes the UAP debate. Rather than a simple binary—extraterrestrial or mundane—it is increasingly a spectrum of possibilities, many rooted in human ingenuity and geopolitical gamesmanship. The electromagnetic battlefield has grown crowded and cunning, filled with signals, decoys, and craft designed to obscure more than reveal. In such an environment, the distinction between unidentified and misidentified is razor-thin—and the ability to discern one from the other has never been more vital.

Thus, while the Tic Tac incident seemed to reveal a nonconventional craft, many subsequent UAP reports may in fact represent misidentified but terrestrial UAVs — often operated without official knowledge, oversight, or attribution.

Mainstream media outlets and streaming platforms capitalized on the UAP frenzy, often reinforcing the most sensational narratives — extraterrestrials, reverse-engineered craft, hidden alien programs. Documentaries, podcasts, and talk shows blurred the line between investigation and entertainment. While these narratives captured public imagination, they also distorted the more grounded national security concerns at the heart of the debate.

At the same time, serious science and serious journalism struggled to find footing. The nuances of propulsion physics, signal analysis, and military telemetry were lost in the noise of social media hype, further polarizing the conversation between skeptics and believers, often to the detriment of evidence-based inquiry.

The aftermath of the Tic Tac revelation laid bare a fundamental vulnerability in the architecture of modern airspace: the absence of a transparent, standardized system for identifying drones—military and civilian alike. In a sky now crowded with unmanned aerial vehicles operated by nation-states, private corporations, hobbyists, and non-state actors, the failure to distinguish between benign and potentially hostile aerial platforms has become more than a bureaucratic shortfall—it is a strategic liability.

The distinction between drone and unidentified aerial phenomenon (UAP) has grown increasingly opaque. An untagged surveillance drone, operating near sensitive military infrastructure, might trigger the same sensor signatures as a genuinely unexplained object. A swarm of UAVs behaving erratically in restricted airspace could be dismissed as a glitch—or recognized too late as a deliberate attempt by a foreign actor to test the limits of U.S. response protocols. Even a commercially available quadcopter, innocuously hovering near a government building, might be harvesting telemetry or conducting targeted reconnaissance on behalf of an adversary.

These hypotheticals were no longer confined to theoretical war games by the time the Chinese high-altitude balloon incidents seized the public's attention during the Biden administration. The image of a foreign surveillance platform floating unchallenged across American territory, only to be brought down after days of political and public pressure, sent shockwaves through the national psyche. It rendered the abstract threat of aerial espionage tangible—visible not just on military radar, but on the nightly news. For many Americans, it raised unsettling questions: If we could not prevent, identify, or decisively respond to such a blatant incursion, what other threats might be passing unseen?

This atmosphere of public anxiety elevated UAP ambiguity from the fringes of defense discourse into the heart of national security policy. The risk was no longer theoretical. With China, in particular, advancing a multi-domain strategy that fuses conventional military strength with cyber capabilities, drone surveillance, and gray-zone operations, the boundaries of American airspace—once assumed inviolable—suddenly seemed porous. The implications were chilling: foreign actors could now challenge U.S. sovereignty with inexpensive, unmanned platforms that hovered somewhere between espionage and provocation, just ambiguous enough to avoid clear retaliatory justification.

Compounding the problem is the lack of international norms or enforceable agreements governing drone behavior across borders. Unlike manned aircraft, which are bound by decades-old conventions and transponder requirements, drones operate in a regulatory gray area—too small, too quiet, and too numerous to track effectively under existing frameworks. Even domestic identification efforts, such as Remote ID mandates in the United States, remain incomplete or unenforced, leaving critical gaps in airspace awareness.

This operational ambiguity does more than complicate identification. It destabilizes command decisions. Military leaders, faced with unidentified objects on radar, must now make split-second judgments without knowing whether they are confronting a hobbyist's misadventure, a spy platform disguised as a weather drone, or something far more advanced and unknown. The cost of a false positive—a downed commercial craft or an international incident—is high. But the cost of a false negative, of inaction in the face of a real threat, could be catastrophic.

Ultimately, the post-Tic Tac world has made one truth unavoidable: in the absence of globally accepted drone classification standards, interoperable tracking systems, and robust attribution mechanisms, the sky will remain a space of strategic ambiguity. And in that ambiguity lies danger—not just from adversaries, but from our own inability to separate signal from noise in a world where surveillance, intrusion, and misidentification are only a few propellers apart.

The Tic Tac incident, and the cascade of UAP reports that followed, did more than ignite public fascination—they cracked open a fault line in the foundation of American air defense and aerospace policy. Long considered the most surveilled and technologically secured airspace on the planet, U.S. military zones were revealed to be porous in ways that no longer aligned with 21st-century threats. Whether these anomalous objects are misidentified drones, clandestine foreign platforms, undisclosed black projects, or technologies that defy current understanding, the implications cut to the heart of national defense and strategic credibility.

At the center of the issue lies a disquieting reality: objects with no transponders, no clear origin, and no discernible propulsion can and have entered restricted military airspace—maneuvering freely, lingering near carrier strike groups, and exiting without confrontation or identification. The 2004 Tic Tac encounter occurred in the midst of a live readiness exercise involving multiple U.S. Navy capital ships, a scenario designed to simulate and stress-test America's warfighting capabilities. And yet, within that fortified operational bubble, an object displaying physics-defying movement was able to operate with impunity. Not intercepted. Not classified. Not explained.

Modern radar and sensor systems, while powerful, are still bound by human assumptions. Software filters are designed to eliminate "anomalous" data—precisely the kind of returns that such objects produce. If a contact moves too fast, shifts direction too abruptly, or reflects no known radar signature, it may be excluded not because it lacks substance, but because it defies expectation. These blind spots, embedded in the very architecture of detection systems, mean that some of the most critical targets may never even register as threats.

Compounding this is the challenge of rules of engagement. In a world where drones can be launched from submarines, repurposed commercial UAVs can conduct espionage, and loitering munitions can impersonate weather balloons, traditional engagement doctrines begin to fracture. What do you do when a radar contact matches no known aircraft, carries no electronic emissions, and exhibits no hostile behavior—but is hovering within range of a nuclear weapons platform? The absence of positive identification and attribution paralyzes action. A strike risks escalation. Hesitation risks vulnerability.

This is not merely an intelligence failure. It is a doctrinal void in how the United States interprets and responds to ambiguous threats in its own skies. And it arrives at a time when adversaries are increasingly investing in platforms designed to exploit precisely this ambiguity. Swarm tactics, signal spoofing, radar jamming, and electromagnetic deception are no longer experimental—they are battlefield realities. The drone age has erased the comfort of visual identification and clearly marked threats. In its place is a contested domain where milliseconds matter, attribution is murky, and overconfidence can be fatal.

The broader implications are strategic. If foreign powers, or unknown actors, can operate undetected and unchallenged above the world's most advanced military formations, then the very premise of air superiority is at risk. Deterrence, after all, is built not only on capability but on perception—on the belief that no threat can cross a red line without consequence. The Tic Tac incident, and the institutional unease it left in its wake, suggests that some of those lines have already been blurred. In doing so, it has exposed a sobering truth: America's dominance in the air may no longer rest on its ability to fly higher or faster, but on its ability to understand, anticipate, and control what it does not yet fully recognize.

One of the most transformative innovations in the evolution of unmanned aerial systems is swarm technology—the ability for multiple drones to operate not as isolated tools, but as a cohesive, intelligent collective. Drawing inspiration from biological systems such as flocks of birds, schools of fish, and insect colonies, drone swarms are revolutionizing both civilian and military applications. Unlike traditional drone operations that rely on direct human control, swarms function through distributed artificial intelligence, real-time mesh communication, and autonomous decision-making, enabling them to adapt dynamically to their environment and objectives.

A defining characteristic of true drone swarms is their decentralized nature. There is no central command drone or lead unit. Instead, each individual drone operates based on shared algorithms and local perception, making decisions autonomously while maintaining coordination with the rest of the swarm. This allows for remarkable self-organization. Swarm members continuously adjust to changing terrain, shifting mission goals, or unexpected obstacles, often without requiring human input. Tasks are distributed based on real-time variables such as location, remaining battery life, or role specialization—whether surveillance, data relay, or payload delivery.

The drones communicate through low-latency mesh networks, sharing their positions, sensor data, and commands across the swarm in milliseconds. This level of interconnectivity gives the swarm the ability to function as a single distributed system, rather than a group of independent units. The result is a system that is adaptive, resilient, and capable of executing complex missions in dynamic environments.

Nowhere is the potential of swarm technology more evident than in military applications. Armed forces around the world are investing heavily in the development of swarm-capable platforms, recognizing their strategic value in the future of warfare. In saturation attacks, large numbers of drones can be launched simultaneously to overwhelm enemy radar and air defense systems, not through stealth or speed, but through sheer volume and erratic behavior. In decoy operations, swarms can mimic the radar signatures of manned aircraft, confusing targeting systems and drawing fire away from primary assets.

Swarms can also serve as persistent reconnaissance webs, maintaining wide-area surveillance with overlapping sensor fields and continuous coverage of contested zones. In complex environments such as urban battlefields or collapsed structures, swarm drones can perform search-and-clear missions, navigating tight spaces, identifying threats, and even relaying vital information back to human operators—all without endangering personnel.

Countries such as the United States, China, Russia, and Turkey are actively pursuing swarm-capable munitions, including AI-guided microdrones and loitering munitions that can identify and engage targets autonomously. These technologies are already shaping the next era of warfare, where drones will not merely support human soldiers but act as independent agents, executing missions with unprecedented speed, scale, and precision. As these capabilities mature, swarm warfare is poised to become not just a tactical advantage, but a foundational element of future military doctrine.

Among the most notable demonstrations of autonomous swarming in the United States was the Perdix Project, a Department of Defense initiative that, in 2016, successfully deployed a swarm of 103 micro-UAVs from fighter jets. These drones operated as a single distributed intelligence, exhibiting collective decision-making, adaptive flight behaviors, and self-healing dynamics—hallmarks of a system no longer dependent on individual control, but functioning as a true autonomous swarm. The success of Perdix marked a turning point in the viability of swarming technology, transforming it from concept to battlefield reality.

Beyond military applications, swarm drones are beginning to reshape the civilian and commercial landscapes. In environmental monitoring, swarms are deployed across forested regions, wetlands, and coastal zones to collect data on deforestation, wildlife movement, or pollution levels. In precision agriculture, multiple UAVs work in tandem—each assigned to monitor a specific crop zone, assess soil moisture, or identify pests—maximizing efficiency while minimizing human labor. During disaster response, swarm drones provide real-time mapping of damage zones, navigate through collapsed structures, and locate survivors far more rapidly than traditional teams. In the realm of entertainment, companies like Intel and SkyMagic have redefined aerial spectacle with synchronized light shows involving hundreds of drones, offering choreographed displays once impossible with pyrotechnics alone.

Logistics and infrastructure sectors are also poised for transformation. Drone swarms are being tested for autonomous warehouse inventory management, parcel tracking in transit, and multi-point delivery systems, where goods are dispatched to various destinations simultaneously with machine-optimized efficiency. In the near future, swarm-enabled logistics could redefine how supply chains operate at scale.

This emergent functionality depends on a convergence of enabling technologies. Modern swarms leverage onboard artificial intelligence chips for navigation, collision avoidance, and mission logic, allowing them to operate independently of constant human input. Simultaneous Localization and Mapping (SLAM) enables drones to explore and navigate unknown terrain while building internal maps on the fly. Computer vision systems, paired with thermal imaging, give drones the ability to detect human figures, recognize landmarks, and identify anomalies in both daylight and obscured conditions. Increasingly, edge computing allows drones to process data locally—reducing reliance on cloud infrastructure and improving responsiveness in bandwidth-limited environments. Integration with 5G networks, and eventually 6G, will enable near-zero latency communication, unlocking the potential for ultra-dense drone operations within cities, industrial complexes, and disaster zones.

Yet with this potential comes significant risk. Cybersecurity vulnerabilities pose serious concerns; drone swarms can be jammed, spoofed, or even hijacked, particularly if communication protocols are unencrypted or insufficiently protected. The rise of lethal autonomous drones has triggered global debate over the ethics of systems that could, theoretically, make life-or-death decisions without human oversight. In the civilian sphere, the challenge of airspace deconfliction—ensuring autonomous drones safely coexist with manned aircraft and each other—remains a critical hurdle to widespread deployment. Compounding these issues is the lack of international standards or legal norms for regulating swarm-capable systems, even as their numbers and sophistication grow rapidly.

Underlying this revolution is a shift in the very nature of the drone. No longer just flying machines, drones are becoming thinking systems—capable of sensing, learning, adapting, and executing complex operations with minimal or no human intervention. As computational power becomes smaller, faster, and more efficient, drones are evolving into fully autonomous agents, able to operate for extended periods in contested or dynamic environments. This transformation is made possible by the fusion of artificial intelligence, machine learning, multi-sensor integration, and real-time decision-making algorithms. The result is not just a smarter aircraft—but a new class of aerial intelligence, capable of navigating both the physical and informational domains with unprecedented autonomy and purpose.

The journey from remote control to autonomy marks one of the most significant transformations in the evolution of unmanned aerial vehicles. In their earliest form, drones were little more than glorified radio-controlled helicopters, entirely dependent on human pilots to guide every movement and decision. Today, however, modern UAVs are capable of remarkable independence. They can chart their own routes using GPS and vision-based mapping, detect and avoid obstacles mid-flight, adapt to weather changes and shifting terrain, identify objects and faces, and modify their behavior in real time based on sensor input and evolving mission needs. This shift from manual control to cognitive autonomy has turned drones from simple tools into intelligent agents—capable of patrolling borders, analyzing traffic, monitoring wildlife, and completing complex missions with minimal human involvement.

At the heart of this transformation are integrated AI systems that allow for multi-layered autonomous behavior. Navigation systems combine GPS with inertial measurement units and visual odometry to determine accurate positioning and pathfinding. Perception systems use LiDAR, radar, thermal imaging, and computer vision to interpret the surrounding environment—recognizing terrain features, buildings, vehicles, and human presence. Decision-making algorithms allow drones to prioritize actions based on changing objectives, real-time inputs, and evolving threat assessments. In swarm deployments, AI enables drones to communicate and coordinate with nearby units, forming a cohesive collective that acts as a single distributed system.

Modern UAVs often operate on modular software platforms such as ROS (Robot Operating System) and PX4, enabling flexible upgrades and mission-specific configurations. These software stacks provide the framework for integrating advanced autonomy, real-time control, and adaptive mission planning, allowing operators to tailor capabilities to specific operational environments.

Across a wide range of domains, autonomous drones are now performing tasks once considered the exclusive domain of human operators. In military reconnaissance, low-profile UAVs scout enemy positions, assess battlefield threats, and gather intelligence far ahead of human deployments. During natural disasters, swarms of drones search for survivors, identify fire hotspots, and deliver aid to remote or inaccessible regions. In infrastructure inspection, drones navigate within confined spaces—inside wind turbines, oil pipelines, or power lines—using advanced vision systems to detect microfractures, corrosion, or structural fatigue. In agriculture, AI-guided UAVs monitor crop health, detect pest infestations, and conduct pinpoint spraying of fertilizers or pesticides with a level of precision that far surpasses traditional methods.

Autonomous drones are even being trained to detect, classify, and track other aerial objects—an emerging need in an increasingly crowded airspace where potential threats can include rogue drones or foreign reconnaissance platforms. These systems allow drones to respond dynamically to uncooperative or unknown aerial contacts, avoiding collisions or engaging in defensive maneuvers if required.

Yet with this power comes a new set of challenges. Predictability remains an issue—AI systems can make decisions that are difficult to explain or anticipate, raising concerns about consistency and safety in unpredictable environments. Accountability is another unresolved dilemma: when an autonomous drone fails, causes harm, or breaches airspace, it is unclear whether the fault lies with the developer, the operator, or the algorithm itself. Most regulatory authorities still mandate some degree of human oversight, especially in or near populated areas. And with growing autonomy comes greater vulnerability: drones must be protected from spoofing, jamming, and cyber intrusion through hardened encryption and resilient software design.

These challenges underscore the enduring importance of **human oversight** in an increasingly autonomous era. Whether through remote pilots, supervisory AI, or embedded ethical constraints within algorithms, human judgment remains essential. As drones evolve from tools to agents, the role of the operator is being redefined—not eliminated. Rather than guiding every flight path, the human becomes a **mission architect, AI trainer, and system supervisor**. The shift is not just technological—it is philosophical. The drone moves from being directed to deciding, from reacting to understanding.

Swarm technology represents the leading edge of this evolution. In these systems, intelligence is no longer centralized in a single vehicle, but distributed across dozens—or hundreds—of units that adapt, coordinate, and execute complex missions as a collective organism. As software improves, sensors become more refined, and regulatory frameworks catch up to the pace of innovation, autonomous aerial collectives may soon become as common over farms, cities, and disaster zones as they are over battlefields. The 21st-century sky is on the brink of a transformation—one shaped not by the number of drones in flight, but by how they think and act together.

Yet as drones grow more agile, more autonomous, and more difficult to detect, they begin to occupy the same ambiguous airspace as unidentified aerial phenomena (UAPs). The boundary between the UAV and the UAP is becoming increasingly porous, introducing strategic complications on multiple fronts. Attribution becomes a challenge; military commanders facing a fast-moving unknown in restricted airspace must make split-second decisions—often without being able to determine whether the object is a commercial drone, a foreign military UAV, or something truly unexplained. Escalation risks rise in kind; a misidentified drone near a nuclear facility or naval group could trigger a diplomatic incident or armed response. Exploitation is also a growing concern. Adversaries may deliberately engineer drones to mimic UAP behavior, using their ambiguity to test U.S. response times, gather intelligence, or sow public and institutional confusion.

The Tic Tac incident—regardless of its ultimate origin—offered a glimpse of how modern airspace ambiguity can confound even the most advanced militaries. It illustrated not just a potential technological gap, but a strategic blind spot. That incident, and the institutional introspection that followed, has since catalyzed a modernization effort across multiple branches of the U.S. government.

In response, the Pentagon and allied agencies are investing in next-generation detection and tracking systems designed to meet the complexity of this new aerial environment. These include artificial intelligence-enhanced radar systems, multi-domain sensor fusion, and spectral signature analysis capable of distinguishing between conventional drone propulsion and anomalous energy outputs. Space-based surveillance platforms are being upgraded to monitor near-space altitudes and track high-speed maneuvering objects that evade traditional radar. Some programs now even incorporate crowdsourced civilian sensor data, selectively fused into broader detection networks to expand coverage and spot patterns that institutional systems alone might miss.

Projects like the NORAD modernization effort, the All-Domain Anomaly Resolution Office (AARO)'s calibration initiative, and DARPA's airspace situational awareness programs all trace part of their urgency to the unanswered questions left in the wake of the 2004 Tic Tac event. They reflect a growing consensus: that drone proliferation and unidentified aerial phenomena are not distinct challenges, but overlapping components of a more complex airspace reality. As drones continue to evolve, the systems that monitor them must evolve faster—because in an era where ambiguity can be weaponized, clarity becomes a form of defense.

The growing ambiguity in global airspace has also elevated the call for international norms and cooperative frameworks to manage drone identification, accountability, and incident reporting. Much like the maritime and space domains—where structured agreements help prevent conflict and promote transparency—the aerial domain is now sufficiently crowded, contested, and complex to demand similar multilateral coordination.

In the aftermath of high-profile encounters and rising drone proliferation, proposals that once seemed overly ambitious are now regarded as practical necessities. Among the most pressing are the establishment of global UAV registries, allowing nations to track and identify authorized unmanned systems across borders. Encrypted transponder protocols, akin to modern aviation identification systems but adapted for autonomous and micro-UAV platforms, are also under consideration to prevent spoofing and unauthorized incursions. Additionally, the creation of shared anomaly data networks among allied countries could provide early warning, improve attribution, and reduce the number of incidents classified as truly unexplained.

These measures aim not only to deconflict civilian, commercial, and military drone activity, but to narrow the gap between the known and the unknown—ensuring that unidentified aerial encounters become less a function of tracking failure and more a true exception worth investigating. As drones continue to reshape the sky, building international trust and operational transparency may become as important as the technology itself.

The 2004 Tic Tac encounter forced a reckoning not only within the halls of the Pentagon, but also across the intelligence community, the scientific world, and the global defense establishment. It presented an object — real, recorded, and witnessed — that could not be readily explained or dismissed. In doing so, it shattered the longstanding complacency surrounding unidentified aerial phenomena and revealed profound gaps in our awareness, classification systems, and technological assumptions.

Crucially, the incident did not reveal aliens, time travelers, or metaphysical forces. Instead, it revealed how ill-prepared the world's most sophisticated military apparatus was to categorize and respond to an unidentified object maneuvering freely within restricted airspace. The problem was not necessarily that the object came from beyond Earth — it was that no one knew what it was, who controlled it, or how it worked. In an age of rapid drone evolution and geopolitical tension, that ignorance was unacceptable.

As a result, the United States entered a new era — one where UAV transparency and airspace accountability became national imperatives. No longer could black programs operate without proper deconfliction protocols. No longer could foreign or domestic drones be dismissed as minor nuisances. And no longer could truly unknown objects be swept under the rug of bureaucratic indifference.

Whether dealing with classified reconnaissance vehicles, commercial quadcopters, or the elusive category of genuine UAPs, the need for clear reporting, unified tracking systems, and cross-agency coordination is now seen as critical to airspace safety and national defense. The Tic Tac encounter was the signal — the response is still unfolding.

In this evolving landscape, the drone is not just a tool — it is a test. It tests the limits of our technologies, the strength of our detection networks, the flexibility of our policies, and the openness of our governments. And when a drone — or something that resembles one — operates outside those boundaries, it demands answers. Not just from the military, but from the public, from scientists, and from the international community.

The Tic Tac incident will be remembered not only for what it showed in the sky, but for what it exposed on the ground: a world at the edge of its own understanding, confronting technologies it may not yet fully grasp. In that moment, the age of drones and the age of aerial anomalies became one and the same.

The question now is no longer "What was it?"

It is: Are we ready for what comes next?

Chapter 15 - Beneath the Surface – The Rise of Unmanned Underwater Vehicles

While aerial drones dominate headlines, a quiet but equally transformative revolution has been unfolding beneath the oceans. Unmanned Underwater Vehicles (UUVs)—often referred to as underwater drones—are reshaping the way we explore, study, and defend the maritime domain. From oceanographic research and infrastructure inspection to covert surveillance and naval warfare, these submersible platforms are expanding the frontier of unmanned systems into one of the most inaccessible and strategically vital environments on Earth.

UUVs are generally divided into two broad categories. Remotely Operated Underwater Vehicles (ROUVs) remain tethered to surface ships or human operators via fiber-optic cables. These systems offer real-time video feedback and allow for precise manipulation of tools, grippers, and sensors. ROUVs are indispensable for delicate or high-stakes tasks, such as undersea cable repair, submarine rescue, and detailed inspection of deep-sea installations.

In contrast, Autonomous Underwater Vehicles (AUVs) operate independently of direct human control. Equipped with pre-programmed mission logic or adaptive artificial intelligence, these platforms are capable of executing complex operations such as seafloor mapping, current profiling, biological sampling, or covert reconnaissance across hundreds or even thousands of nautical miles. AUVs rely on internal navigation systems, sonar, and inertial guidance to function where GPS and radio signals cannot penetrate.

Modern UUVs come in a wide array of shapes, sizes, and propulsion systems—from biomimetic miniatures, designed to mimic fish or squid for close-proximity research, to nuclear-powered torpedo-length vehicles capable of crossing entire ocean basins and remaining submerged for months at a time.

The development of UUVs traces back more than half a century. In 1957, the Special Purpose Underwater Research Vehicle (SPURV), developed at the University of Washington, became the first operational AUV, designed for Arctic scientific missions. Through the 1970s to 1990s, advancements accelerated: solar-powered models, longer endurance thrusters, and increasingly robust sensor arrays expanded both civilian and military applications. Soviet models like the L1 and L2, and U.S. Navy programs such as ARCS (Autonomous and Remotely Controlled Submersible), began to test the outer limits of underwater autonomy.

By the early 2000s, UUVs moved beyond the laboratory and entered commercial service. The offshore oil and gas industries began deploying them for pipeline inspection, mineral prospecting, and deep-sea infrastructure monitoring. Simultaneously, national navies began to take note of their potential for mine countermeasures, undersea surveillance, and anti-submarine warfare.

By the 2020s, UUVs crossed into a new frontier: weaponization. Platforms such as Ukraine's Toloka TLK-150 and DARPA's Manta Ray have introduced features such as stealth shaping, seafloor energy harvesting, and autonomous target identification. These systems can navigate complex undersea terrain, evade sonar detection, and carry out precision strikes—all without a direct link to their human operators. The underwater battlespace, long dominated by manned submarines and stationary sensors, is now becoming a dynamic, machine-driven domain.

What began as an academic tool for polar research has evolved into one of the most versatile and strategically significant classes of unmanned systems. As UUVs become faster, smarter, and more self-sufficient, they are poised to change the rules of undersea engagement, redefine maritime domain awareness, and reveal parts of our planet still unexplored or long forgotten beneath the waves.

The militarization of Unmanned Underwater Vehicles (UUVs) has accelerated rapidly, transforming them from passive reconnaissance platforms into stealthy, multi-role assets capable of reshaping naval warfare. Their utility spans the spectrum of maritime operations. In mine detection and clearance, small autonomous systems like the REMUS AUV have demonstrated the ability to neutralize threats across underwater minefields in a matter of hours—a task that, for human divers, could take weeks and expose personnel to extreme risk.

For surveillance and reconnaissance, UUVs offer long-endurance, low-visibility monitoring of contested waters. China's Sea Wing gliders have been regularly spotted in the South China Sea, quietly patrolling sensitive zones under the guise of oceanographic research. Meanwhile, Russia has developed the Poseidon nuclear-powered UUV, capable of delivering a thermonuclear payload across oceans autonomously, underscoring the high-stakes strategic value now associated with undersea platforms.

In combat operations, the U.S. Navy and its defense contractors are advancing a new generation of armed, loitering UUVs with extended range, autonomous navigation, and stealth technologies. One of the most recent breakthroughs came in April 2025, when Anduril Industries unveiled the Copperhead—a family of autonomous underwater systems designed for high-endurance naval missions, including threat interdiction, decoy deployment, and mine warfare.

These advancements have been enabled by a wave of design innovations. The Manta Ray UUV, developed independently by both China and Northrop Grumman, mimics the shape and movement of marine animals to reduce sonar signatures and enhance hydrodynamics. Glider drones, which shift buoyancy instead of relying on traditional propellers, can silently traverse long distances, making them ideal for submarine detection and deep-sea research. New propulsion technologies, including lithium–water and hydrogen fuel cells, have significantly increased endurance, allowing UUVs to remain submerged and operational for weeks or even months. The modular construction of many platforms enables rapid deployment, including airdrop delivery or field assembly in operational theaters. In 2024, DARPA's Manta Ray prototype successfully completed open-sea trials, demonstrating stealth, energy harvesting from the ocean, and reconfigurable architecture for multiple mission profiles.

Beyond warfare, UUVs play a vital role in scientific discovery and environmental exploration. Autonomous platforms are now exploring the ocean's deepest trenches, collecting samples of unknown life forms, locating hydrothermal vents, and generating high-resolution maps of previously uncharted seafloor. The Sentry AUV, developed by Woods Hole Oceanographic Institution, can operate at depths of 6,000 meters, while Argo, an earlier platform, played a central role in locating the wreck of the Titanic and providing footage for National Geographic's Secrets of the Titanic. In 2021, a class of bio-inspired soft robots demonstrated survivability at Mariana Trench depths, introducing a new generation of flexible, pressure-resistant UUVs.

Civilian uses are also expanding. In ecosystem rehabilitation, UUVs monitor water quality in urban waterways, contributing to restoration projects in places like New York City's Bronx River and Harlem Estuary. For climate monitoring, swarms of glider drones track shifts in temperature, ocean acidity, and current dynamics, providing critical input for climate models. The film and documentary world has also embraced UUVs for underwater cinematography, capturing high-definition visuals of wrecks like the Lusitania and Titanic from previously unreachable angles.

Yet operating underwater presents unique communication and control challenges. Unlike aerial drones, UUVs cannot rely on GPS or standard radio links. Underwater communications must depend on acoustic transmissions, which suffer from delay, distortion, and loss due to refraction, reflection, and terrain interference. Even in ideal conditions, latency can range from one to two seconds, and signals can degrade significantly in cluttered environments. These constraints necessitate high levels of autonomy—UUVs must often execute entire missions with pre-loaded logic and limited or delayed opportunities for human intervention. This autonomy is both an advantage and a liability: it enables stealth and persistence but also introduces a security paradox. Once deployed, a UUV may operate beyond meaningful human oversight, raising concerns about accountability and command control.

The strategic implications of these platforms are profound. Capable of tracking submarines, laying mines, or launching kinetic strikes, UUVs can operate undetected in contested maritime regions, making them highly destabilizing tools in modern naval doctrine. According to the RAND Corporation, UUVs now represent a key pillar in future strategies for submarine warfare, anti-access/area-denial (A2/AD), and tactical surprise. China's pursuit of flying submarine drones—platforms that can transition between aerial and undersea operation—demonstrates that the next arms race may center not just on advanced firepower, but on multi-domain autonomy.

As UUVs continue to push the limits of endurance, stealth, and decision-making independence, the oceans may become the most contested—and least visible—theater of unmanned evolution.

Chapter 16: Conclusion – Navigating a Drone-Filled World

The ascent of drones into mainstream society marks one of the most rapid technological integrations in modern history. What began as experimental military craft has evolved into a multifaceted technology that touches nearly every sector—defense, commerce, healthcare, journalism, agriculture, infrastructure, and beyond. As UAVs become more autonomous, interconnected, and indispensable, the urgency to govern their rise responsibly grows ever more critical.

We now stand at a crossroads where policy must catch up with innovation, ethics must guide capability, and collaboration must override competition.

The pace of drone innovation has routinely outstripped the ability of governments to regulate it. Agencies around the world are racing to adapt:

The Federal Aviation Administration (FAA) has led domestic efforts through its Part 107 regulations, establishing clear parameters for commercial drone operations. These include mandatory certification, altitude restrictions, visual line-of-sight (VLOS) requirements, and prohibited zones near airports and critical infrastructure.

Yet the frontier remains unsettled. Beyond Visual Line of Sight (BVLOS) operations—crucial for long-range delivery, infrastructure monitoring, and UAM—are only permitted through special waivers. The FAA's anticipated rulemaking in 2025 aims to formalize BVLOS protocols, supported by detect-and-avoid technologies and remote identification systems.

Globally, the International Civil Aviation Organization (ICAO) has been instrumental in promoting standardized UAV integration into controlled airspace. It supports national regulators in developing harmonized safety and communication standards. Similarly, the European Union Aviation Safety Agency (EASA) offers a model framework with its three-tier drone categorization: Open, Specific, and Certified, allowing tailored regulation based on operational risk.

Still, a patchwork of national rules hampers cross-border drone operations—an obstacle to future global applications such as international air cargo, climate monitoring, and disaster relief coordination.

As drones proliferate across civilian and military domains, a parallel discourse has emerged—one that grapples with the balance between **technological capability and democratic accountability**. At the heart of this debate lies a central tension: how can societies embrace the transformative benefits of drone technology without undermining **civil liberties, international norms, or ethical guardrails**?

Nowhere is this tension more visible than in the realm of **surveillance and privacy**. Drones equipped with **high-resolution cameras**, **thermal sensors**, and **facial recognition software** have granted law enforcement agencies a form of **aerial omnipresence** once reserved for military intelligence. In urban centers, these systems have been deployed to **monitor protests**, **track movement patterns**, and **enforce public order**. But their presence has also raised alarms. Civil liberties organizations, including the American Civil Liberties Union (ACLU), have issued strong calls for warrant requirements, data retention limits, and transparent usage logs. The concern is not only about potential abuse—but also about the chilling effect such surveillance might have on free speech and public assembly, cornerstones of democratic society.

In the military domain, drones have revolutionized modern warfare—enabling surgical strikes, reducing risk to human forces, and extending reach deep into hostile or inaccessible territory. But these advantages have come at a cost. Programs such as the Disposition Matrix, which guides drone-based targeted killings, have faced intense criticism for lack of transparency, civilian casualties, and due process violations. The 2021 Kabul drone strike, which mistakenly killed ten civilians—including children—underscored the profound risks of flawed intelligence, algorithmic decision-making, and overreliance on remote assessments. Globally, the deployment of armed drones in sovereign airspace without consent continues to challenge international law, while non-state actors, from insurgent groups to terrorist networks, increasingly deploy drones for asymmetric warfare. The rapid development of lethal autonomous weapons systems, often ahead of any meaningful regulatory framework, only amplifies the moral and legal urgency.

These challenges are not confined within borders. In an age where drones can cross jurisdictions, enter restricted airspace, or gather intelligence from foreign shores with ease, there is an urgent need for global governance. Like maritime and space law before it, aerial governance must evolve to address the unique risks and responsibilities of unmanned systems.

Efforts toward a unified framework are gaining momentum. Key international institutions—including the International Civil Aviation Organization (ICAO) and the United Nations Office for Disarmament Affairs (UNODA)—have begun exploring global norms for remote identification, airspace deconfliction, and data transmission protocols. A multilateral system of accountability, supported by shared regulatory standards, is critical to prevent miscommunication, ensure public safety, and uphold sovereignty.

Ethical considerations must also be codified. Just as chemical and biological weapons are subject to strict international prohibitions, so too must there be conventions governing the ethical deployment of autonomous drones, civilian privacy protections, and the use of persistent aerial surveillance. Institutions and governments must agree on where the line lies between innovation and intrusion.

Transparency will be equally essential. Publicly accessible tools like the Global Drone Regulations Database offer a foundation for international cooperation and civilian awareness. Expanding such platforms to include mission reporting, impact assessments, and cross-border incident tracking could build trust and preempt crises. In parallel, there is growing consensus on the need for technology transfer controls—to prevent the proliferation of advanced drone capabilities to authoritarian regimes or non-state combatants. A global export control regime—akin to the Wassenaar Arrangement for dual-use technologies—may be necessary to curb destabilizing deployments.

Looking ahead, navigating the drone age will require more than innovation—it will demand vision. A future built on unmanned systems must rest on a foundation of law, accountability, and public trust. National airspace systems must be modernized to accommodate both manned and unmanned vehicles safely. Citizens must be educated about their rights, responsibilities, and vulnerabilities in a world filled with flying sensors and semi-autonomous actors. And above all, drones must be designed for trust—with safeguards, transparency, and ethical logic embedded into every layer of their operation.

The potential benefits are vast. Faster delivery of goods and aid. Smarter urban planning. More effective environmental monitoring. Sharper military precision. But the risks—from algorithmic bias to privacy erosion, cyber sabotage, and automated violence—are just as real. If managed wisely, drones could become cornerstones of resilience, equity, and global innovation. If neglected, they may become instruments of division, surveillance, and disruption.

The sky is no longer the limit. But how we choose to navigate it will define the trajectory of this century.

Epilogue – The Multidomain Drone Future

The era of unmanned systems began quietly—marked by hesitant experiments, improvised prototypes, and battlefield curiosities. At first, they were side notes to human ingenuity: awkward, glitch-prone machines that hinted at future promise. But over time, the promise became reality. Today, uncrewed vehicles dominate every domain of modern conflict and commerce, extending human reach into skies, seas, cities, and subterranean depths.

This transformation has unfolded across three primary domains, each contributing a distinct yet interconnected revolution. In the air, Unmanned Aerial Vehicles (UAVs) now patrol borders, deliver consumer goods, and carry out precision strikes. The same skies that once held only manned aircraft are now populated by Bayraktar TB2s over Ukraine, MQ-9 Reapers over Syria, and Amazon drones over suburbia—each a signal that rotorcraft and algorithms are now as decisive as pilots and planes.

On the surface of the sea, Uncrewed Surface Vessels (USVs) monitor shipping lanes, gather intelligence across contested chokepoints like the Strait of Hormuz, and serve as swarming decoys in emerging maritime strategies. The Black Sea has become a proving ground for their disruptive potential, showing how unmanned vessels can shape naval operations without a single crew aboard.

Beneath the waves, Uncrewed Underwater Vehicles (UUVs) operate in a realm of silence and shadow. These machines detect undersea mines, monitor submarine movement, explore the abyssal plains, and prepare for strategic missions that might one day unfold beyond human reach. The ocean, long a sanctuary for stealth, is becoming a domain of autonomous surveillance and influence.

Each platform—whether flying, floating, or submerged—represents a node in a much larger transformation: the decentralization of power, the rise of machine autonomy, and the automation of presence. Together, they are reshaping not only military doctrine, but also how societies think about infrastructure, mobility, and control.

This convergence is changing the very nature of how machines operate. Mosaic warfare, manned-unmanned teaming (MUM-T), and cross-domain operations are no longer abstract theories—they are operational imperatives. The lines between air, land, sea, and space are blurring into a single, integrated ecosystem. Drones now communicate with satellites in real time. AI algorithms process sonar pings from ocean-floor sensors and dispatch aerial drones to investigate anomalies. Undersea gliders lay surveillance arrays that cue surface drones to broadcast findings through UAV relays overhead.

This is not merely a shift in platforms. It is a shift toward networked autonomy, a world where the effectiveness of any one system depends not on its hardware, but on its ability to share, adapt, and decide. Autonomy in this context is no longer about independence—it is about interdependence.

Yet, with this power comes a host of new paradoxes. Responsibility becomes diffuse. When a drone fires a missile based on algorithmic judgment, who is accountable? The coder? The commander? The machine? Security becomes fragile; autonomous systems are vulnerable to hacking, spoofing, jamming, and electromagnetic disruption. A single exploited vulnerability could cascade across a networked swarm. And escalation becomes more likely. A misunderstanding between machines, a misread signal, or a spoofed transmission could trigger conflict faster than any diplomat can respond.

As these systems proliferate, the balance of global power is no longer measured solely by industrial capacity or troop strength—but by software architecture, sensor fidelity, and machine decision logic. Code is becoming as strategically important as steel.

Looking to the future, smart cities will fuse air, sea, and underground drone infrastructure, where urban package delivery coexists with underwater waste monitoring and rooftop vertiports. Navies will field unmanned squadrons that operate autonomously in every ocean, from the Arctic ice floes to the coral archipelagos of the Pacific. Corporations will entrust logistics, inspection, and environmental sensing to fleets of tireless, precise, and rapidly scalable autonomous agents.

But the greatest challenge will not be technological. It will be governance. As borders become porous to uncrewed machines and autonomous agents operate beyond traditional jurisdictions, humanity will be forced to answer fundamental questions: How do we ensure accountability in a machine-run world? How do we preserve human dignity in the age of autonomous surveillance? How do we balance innovation with restraint?

The drone age promises immense benefits—efficiency, precision, reach, and resilience. But left unchecked, it also poses existential threats to civil liberties, global stability, and the human role in decision-making. If guided wisely, drones can be instruments of equity, protection, and progress. If not, they may become tools of coercion, opacity, and uncontrolled escalation.

The sky is no longer the limit. The ocean is no longer unreachable. The battlefield is no longer human-only. But how we chart our course through this new frontier—above, below, and beyond—will define the character of our civilization in the century to come.

Glossary of Terms

AI (Artificial Intelligence) – Computer systems designed to simulate human intelligence, enabling drones to navigate, make decisions, and execute tasks autonomously.

Anduril – A defense technology company developing autonomous systems, including the YFQ-44A Fury drone for the USAF's Collaborative Combat Aircraft program.

Autonomous Drone – An unmanned aerial vehicle capable of operating without real-time human control, using onboard sensors and AI for navigation and decision-making.

Bayraktar TB2 – A medium-altitude, long-endurance UAV developed by Turkey's Baykar Defense, widely used in conflicts including Ukraine, Libya, and Nagorno-Karabakh.

BVLOS (Beyond Visual Line of Sight) – Drone operations conducted without the operator having visual contact with the aircraft, requiring advanced communication and detection systems.

CCA (Collaborative Combat Aircraft) – A U.S. Air Force initiative to develop autonomous UAVs that operate alongside manned aircraft to enhance combat capability.

Counter-UAS (C-UAS) – Technologies and strategies designed to detect, track, and neutralize unauthorized or hostile drones.

DJI – Da-Jiang Innovations, a Chinese drone manufacturer that dominates the global civilian and commercial drone market.

Drone Swarm – A group of UAVs that operate collectively under a single command structure or using distributed AI to perform coordinated missions.

eVTOL (Electric Vertical Takeoff and Landing) – Aircraft that use electric power for vertical takeoff, hover, and landing, intended for urban air mobility applications.

FPV (First Person View) – A drone piloting method where operators see through the drone's camera feed in real time, commonly used in racing and tactical operations.

General Atomics – U.S. defense contractor known for developing the Predator, Reaper, and YFQ-42A drones used in surveillance and strike operations.

ISR (Intelligence, Surveillance, and Reconnaissance) – Military operations that gather information to support decision-making and targeting.

Loitering Munition – A drone designed to hover in a target area before identifying and attacking a target, combining the features of a UAV and a missile.

MQ-1 Predator – A foundational U.S. drone used in surveillance and targeted strike missions since the 1990s, retired in 2018.

MQ-9 Reaper – A successor to the Predator drone, offering greater payload, endurance, and versatility for ISR and precision strike missions.

Part 107 – FAA regulations governing commercial drone operations in the U.S., including pilot certification and operational limitations.

Shahed-136 – An Iranian loitering munition drone used extensively by Russia in Ukraine, rebranded as Geran-2.

Swarm Technology – Coordinated control of multiple UAVs that communicate and operate as a collective unit to perform complex tasks.

Unmanned Aerial Vehicle (UAV) – An aircraft without a human pilot on board, remotely controlled or autonomously operated.

Historical Timeline: Key Milestones in Drone Evolution

This timeline captures the pivotal moments, programs, and breakthroughs in the development and deployment of unmanned aerial vehicles (UAVs), from early concepts to today's autonomous battlefield platforms.

19th Century – Early Concepts

· 1849 – Austria uses pilotless balloons loaded with bombs to attack Venice—considered the first recorded use of unmanned aerial weapons.

Early 20th Century – Foundations of UAV Warfare

· 1916–1918 – U.S. develops the Kettering Bug, an early cruise missile guided by gyroscope—predecessor of modern drones.

· 1935 – UK introduces the DH.82B Queen Bee, a radio-controlled target drone for training anti-aircraft gunners.

· 1939–1945 (World War II):

o Germany deploys the V-1 Flying Bomb, the first mass-produced cruise missile.

o U.S. develops Project Aphrodite, converting B-17s into explosive-laden remote-controlled bombers.

o Radioplane OQ-2 becomes the first mass-produced UAV in the U.S.

Cold War Era – Strategic Surveillance Emerges

· 1950s–1960s – U.S. develops the Ryan Firebee, used for reconnaissance over Vietnam and China.

· 1962 – CIA's U-2 spy plane incident over Soviet Union highlights need for unmanned reconnaissance.

· 1964–1971 – Development of the Lockheed D-21, a Mach 3+ high-altitude reconnaissance drone launched from the M-21 variant of the A-12 OXCART program.

· 1970s – Israel pioneers battlefield drones, using real-time video feed in the Yom Kippur War.

1980s–1990s – Transition to Tactical and Armed Drones

· 1982 – Israel uses Scout and Mastiff drones in the Lebanon War—an early example of integrated battlefield ISR.

· 1995 – The MQ-1 Predator makes its first flight, marking the transition to long-endurance UAV surveillance.

· 1999 – Predator used by NATO over Kosovo during Operation Allied Force.

· 2001 – Predator is weaponized with Hellfire missiles; drone strike capability becomes operational.

Post-9/11 Era – The Drone Becomes a Weapon

· 2002 – First known CIA drone strike in Yemen kills al-Qaeda operatives—marking the start of targeted killings.

· 2006–2015 – Massive expansion of U.S. drone operations in Afghanistan, Pakistan, Yemen, Somalia, Iraq, and more.

· 2010s – Proliferation of UAVs globally; China, Turkey, and Iran enter the drone export market.

2015–2020 – Loitering Munitions and Drone Swarms

· 2015 – Turkey's Bayraktar TB2 enters service.

· 2019 – Iran uses drones and missiles to attack Saudi Aramco oil facilities, disrupting global oil production.

· 2020 – Azerbaijan's use of Bayraktar TB2s and Israeli Harop drones decimates Armenian forces in Nagorno-Karabakh War.

2022–2024 – Drone Warfare in Ukraine

· Feb 2022 – Russia invades Ukraine; drones play central role in conflict.

o Ukraine uses Bayraktar TB2s, DJI drones, and FPVs for both surveillance and strikes.

o Russia deploys Orlan-10, Lancet, and Shahed-136/Geran-2 loitering drones from Iran.

· 2023 – Extensive drone-on-drone warfare; both sides conduct daily UAV operations for recon and attack.

2025 – Rise of Autonomous Combat Drones

· March 2025 – USAF designates first CCA platforms as YFQ-42A (General Atomics) and YFQ-44A (Anduril).

· May 2025 – Ground testing begins for both platforms under the Collaborative Combat Aircraft (CCA) program.

· 2025 and beyond – Drones equipped with AI, swarming, and eVTOL technologies begin transitioning into military, civilian, and urban infrastructure roles.

Key Milestones in Drone Evolution

15th Century – Early Concepts

1949. Austria, Piltertsopoụ,hö ạdions loaded with bombs are used in ancı

11990s-1990s. Suit-CAmtidate drons used by Israet in conflict

1982, CIA's U-2 spy plane şụrs UAV reconnasisance

Early 20th Century – Foundaflons of UAV Warfare

1925, UK Asploys DH, 82B Queen.B drones as aerial targets

1989-1945, Lockhee D.21 driving aircraft – Project Aphrodite

Cold War Era – Transition to Tactical and Armed Drones

1982, Scout and Mastiff droness used by Israel in Lebanon

1985, MQ-1 Prodato drones first flight

1980s–1990s – Transitict o Tactal and Armed Drones

1982, U.S. Ryan Firobee drones used in Lebanon war

1935, MQ-1 Predator vodes first fiight

Post-9/11 Era – The Drone Becomes a Weapon

2002, First CIA dronetrike in Yemen

2006-2015 Expanded U.S. drone operations expr

2015–2020 – Loitering Munitions End Drone Svarms

2015, Turkey's Bayraktar TB2 dronos in service

2015, Iran Afta Saudi Aramco li faciilties an Nagoryno-Karaba' war

2022-2024 – Loitering Munitions and Drone Svarms

2015, Turkey's Bayraktar TB2 drones un to concr in

2019, Iran attacc Saudi Arantao in the Nagorho-Karababah War

2022-2024– Drone Warfare in Ukraine

March 2025, U S. Air Force designates first CCA platforms YFO-42A and YFQ-44A and YFQ-44A

Duoses with AI, swarming, and éVTOL transition

into military, civilian, uitas inuto th infrastructure roles

2025 and bey onad

March 2025. Rise of autonomos combat drones

2025 and beyond

March 2025. SA⸢ ȷnat first CCA platforms YFQ-42A and YFQ-44A, PJA

Ground testing begms

Drones with AI, swàrming, ànd eVTOL transition into military.civilla,n/, urban intrastructure reles

2025– Rise of utonomous Combat Drones

About the Author:

Thornton D. "TD" Barnes is a distinguished author, entrepreneur, and former military intelligence specialist whose life's work bridges Cold War espionage, advanced aerospace development, and political commentary. Born in Dalhart, Texas, and raised on a ranch near Clayton, New Mexico, Barnes entered public service early, graduating high school in Oklahoma before enlisting in the U.S. Army. He spent a decade in uniform, specializing in intelligence operations, missile systems, and radar electronics during the height of the Soviet threat.

After his military career, Barnes became a key figure in America's top-secret aerospace initiatives. He supported the CIA's A-12 Project OXCART and stealth programs at Area 51, as well as NASA's X-15 rocket plane, the NERVA nuclear rocket, and five atomic bomb tests at the Nevada Test Site. He later contributed to spaceflight programs from the civilian side, including work with Apollo-era capsule systems while stationed at Wright-Patterson Air Force Base.

Politically, Barnes served as a delegate to the Oklahoma Republican Convention (1976) and Colorado Republican Convention (1980), gaining firsthand insight into the nation's shifting ideological landscape. In civilian life, he led a successful oil and gas exploration company for over four decades, also venturing into uranium and gold mining.

Barnes is the longtime president of Roadrunners Internationale, an association of former CIA and Air Force personnel from Area 51, and serves as Director Emeritus of the Nevada Aerospace Hall of Fame. His contributions have been recognized in documentaries on National Geographic, the History Channel, and numerous intelligence and aerospace publications.

His published works include The Secret Genesis of Area 51, CIA Station D – Area 51, and the Space: The Final Frontier series, along with multiple historical and speculative fiction titles, including The EMP Survival Series and The UFO Phenomenon.

Barnes's most recent nonfiction includes The Woke Movement, Sanctuary Nation, The Russia Gate Conspiracy, and The Hunter Biden Laptop — part of his sweeping political series, The Potomac Swamp, chronicling America's cultural and ideological flashpoints in the post-Trump era.

He resides in Henderson, Nevada, where he continues to write, speak, and consult on topics ranging from intelligence history to the future of American political identity.

Appendix

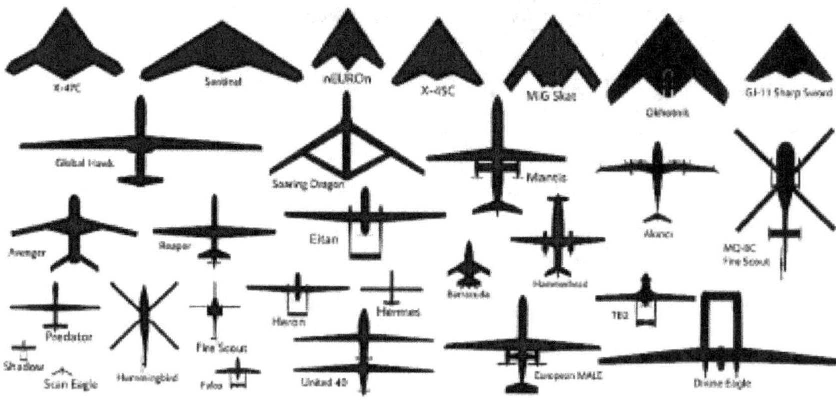
Sentinel
nEUROn
X-45C
MiG Skat
GJ-11 Sharp Sword
Global Hawk
Soaring Dragon
Mantis
Avenger
Reaper
Eitan
MQ-8C
Fire Scout
Barracuda
Heron
Hermes
TB2
Predator
Fire Scout
Shadow
Scan Eagle
Hummingbird
United 40
European MALE
Divine Eagle

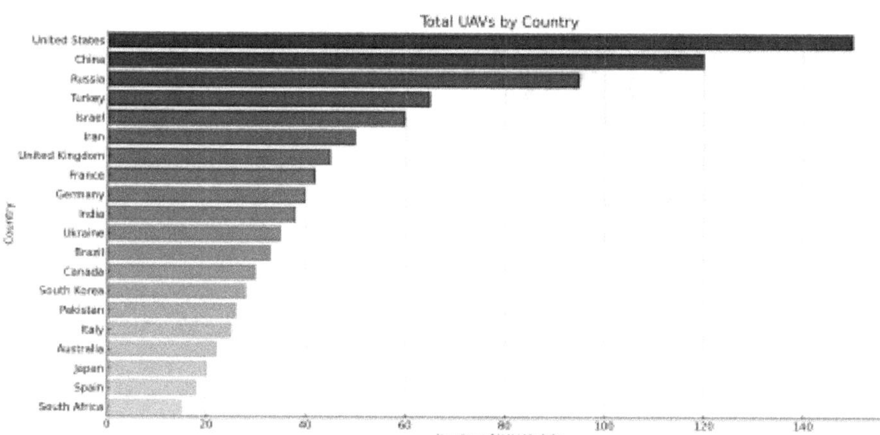
Total UAVs by Country
United States
China
Russia
Turkey
Israel
Iran
United Kingdom
France
Germany
India
Ukraine
Brazil
Canada
South Korea
Pakistan
Italy
Australia
Japan
Spain
South Africa
Country
0
20
40
60
80
100
120
140
Number of UAV Models

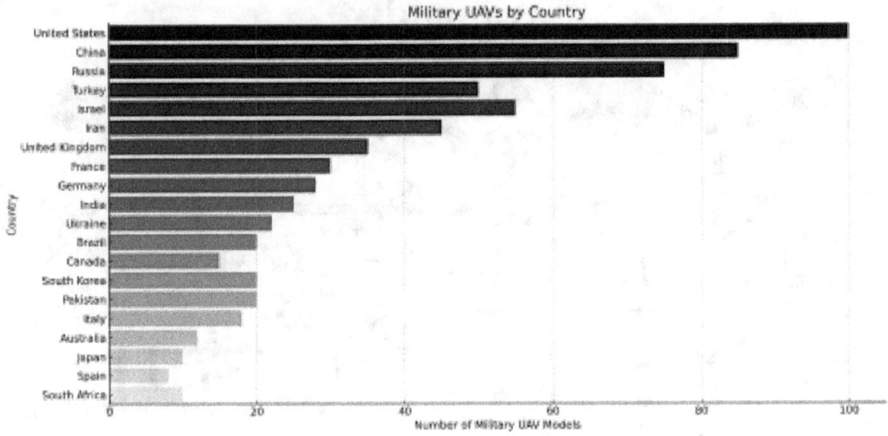
Military UAVs by Country
United States
China
Russia
Turkey
Israel
Iran
United Kingdom
France
Germany
India
Ukraine
Brazil
Canada
South Korea
Pakistan
Italy
Australia
Japan
Spain
South Africa
Country
0
20
40
60
80
100
Number of Military UAV Models

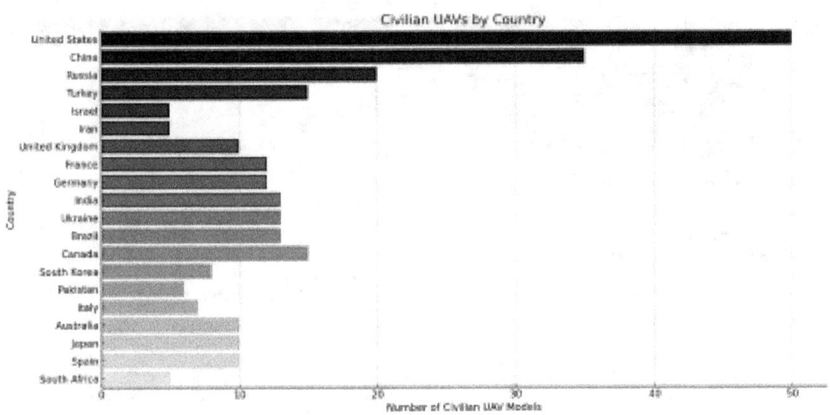
Civilian UAVs by Country
United States
China
Russia
Turkey
Israel
Iran
United Kingdom
France
Germany
India
Ukraine
Brazil
Canada
South Korea
Pakistan
Italy
Australia
Japan
Spain
South Africa
Country
0
10
20
30
40
50
Number of Civilian UAV Models

Bibliography

This bibliography includes books, official documents, reports, and articles referenced in the development of *Drone Evolution*:

Books

· Barnes, Thornton D. *CIA Station D – Area 51*. Begell House, 2023.

· Whittle, Richard. *Predator: The Secret Origins of the Drone Revolution*. Henry Holt, 2014.

· Singer, P.W. *Wired for War: The Robotics Revolution and Conflict in the 21st Century*. Penguin, 2009.

Government and Defense Sources

· U.S. Department of Defense. *Unmanned Systems Integrated Roadmap FY2013–2038*. Office of the Under Secretary of Defense for Acquisition, Technology and Logistics.

· Federal Aviation Administration. *Part 107 – Small UAS Rule*. 2016.

· Air Force Research Laboratory. *Collaborative Combat Aircraft Program Overview*. 2024–2025.

Reports and White Papers

· RAND Corporation. *The Future of Drone Warfare*. 2021.

· CSIS. *Drones and the Future of Air Warfare in the Middle East*. Center for Strategic & International Studies, 2022.

· Brookings Institution. *Civilian Drones, Privacy, and the Future of Airspace Governance*. 2020.

News and Online Media

· The War Zone (The Drive). "Our First Look At YFQ-44A Fighter Drone Prototype." May 2025.

· Breaking Defense. "Air Force Picks Beale AFB to Host CCA Drone Wingmen." May 2025.

· Aviationist. "YFQ-42A and YFQ-44A: The U.S. Air Force's First Fighter Drone Designations." March 2025.

· Flight Global. "USAF Designates First Uncrewed Fighter Prototypes." 2025.

· Reuters. "Iranian Drones Used in Russian Attacks on Ukraine." October 2022.

· CNN. "Drone Swarms Threaten U.S. Navy Vessels." July 2021.

· BBC. "How Drones Are Changing Warfare in Ukraine." 2023.

Don't miss out!

Visit the website below and you can sign up to receive emails whenever TD Barnes publishes a new book. There's no charge and no obligation.

https://books2read.com/r/B-A-YXRJB-TFLHG

BOOKS 2 READ

Connecting independent readers to independent writers.

www.ingramcontent.com/pod-product-compliance
Lightning Source LLC
LaVergne TN
LVHW052309200926
848481LV00002B/510